HOTTR

56 Advances in Polymer Science

Fortschritte der Hochpolymeren-Forschung

Anionic Polymerization

With Contributions by
L. J. Fetters, J. Lustoň, R. P. Quirk,
F. Vašš, R. N. Young

With 21 Figures and 30 Tables

Springer-Verlag
Berlin Heidelberg New York Tokyo
1984

ISBN-3-540-12792-5 Springer-Verlag Berlin Heidelberg New York Tokyo
ISBN-0-387-12792-5 Springer-Verlag New York Heidelberg Berlin Tokyo

Library of Congress Catalog Card Number 61-642

Typesetting: Th. Müntzer, GDR; Offsetprinting: Br. Hartmann, Berlin; Bookbinding: Lüderitz & Bauer, Berlin

2152/3020–543210

Editors

Editorial

With the publication of Vol. 51, the editors and the publisher would like to take this opportunity to thank authors and readers for their collaboration and their efforts to meet the scientific requirements of this series. We appreciate our authors concern for the progress of Polymer Science and we also welcome the advice and critical comments of our readers.

With the publication of Vol. 51 we should also like to refer to editorial policy: *this series publishes invited, critical review articles of new developments in all areas of Polymer Science in English (authors may naturally also include works of their own)*. The responsible editor, that means the editor who has invited the article, discusses the scope of the review with the author on the basis of a tentative outline which the author is asked to provide. Author and editor are responsible for the scientific quality of the contribution; the editor's name appears at the end of it.
Manuscripts must be submitted, in content, language and form satisfactory, to Springer-Verlag. Figures and formulas should be reproducible. To meet readers' wishes, the publisher adds to each volume a "volume index" which approximately characterizes the content.

Editors and publisher make all efforts to publish the manuscripts as rapidly as possible, i.e., at the maximum, six months after the submission of an accepted paper. This means that contributions from diverse areas of Polymer Science must occasionally be united in one volume. In such cases a "volume index" cannot meet all expectations, but will nevertheless provide more information than a mere volume number.

From Vol. 51 on, each volume contains a subject index.

Editors   Publisher

Table of Contents

Anionic Polymerizations of Non-Polar Monomers Involving Lithium

Ronald N. Young
The Department of Chemistry The University of Sheffield, Sheffield, S3 7HF, GB
Roderic P. Quirk*
Michigan Molecular Institute 1910 St. Andrews Drive, Midland Michigan 48640, U.S.A.
L. J. Fetters**
The Institute of Polymer Science The University of Akron, Akron, Ohio 44325, U.S.A.

This article deals with developments in the field of anionic polymerization of nonpolar monomers by the use of the lithium counter-ion. The topics discussed include structure and bonding of organolithium compounds, the initiation and propagation reactions in hydrocarbon solvents, the influence of polar co-solvents on chain propagation, the stereochemistry of polydienes and chain end functionalization. Furthermore, the bonding energies of organolithium aggregates are discussed with regard to the identity of the active species in the initiation and propagation events.

 * Present Address: The Institute of Polymer Science The University of Akron, Akron, Ohio 44325, U.S.A.
** Present Address: Exxon Research and Engineering Co. Corporate Research-Science Laboratories, Clinton Township, Annandale, N.J. 08801, U.S.A.

Advances in Polymer Science 56
© Springer-Verlag Berlin Heidelberg 1984

1 Introduction

Carbanionic polymerizations involving lithium and non-polar monomers have achieved a position of special interest and importance as a result of the potential for obtaining systems lacking spontaneous termination reactions; a feature which was first recognized by Ziegler [1]. The non-terminating nature of these systems facilitates kinetic studies, the preparation of polymers of narrow molecular weight distributions and predictable molecular weights, the synthesis of block copolymers of uniform composition and molecular weight, and allows controlled termination reactions where star-shaped or comb-type polymers can be formed as well as chains having functional groups at one or both ends. Furthermore, polydienes of high 1,4-content can be prepared in hydrocarbon media while the microstructure can be varied by the addition of modifiers to the polymerization system, e.g., the use of bis-piperidino ethane leads to the preparation of poly(vinyl ethylene) from 1,3-butadiene [2].

Thus, even though limited to relatively few monomers, anionic polymerizations exhibiting the foregoing features have attracted both academic and commercial interest. This review covers various aspects of these systems.

2 Structure and Bonding in Organolithium Compounds

Organolithium compounds are unique among the organic derivatives of the alkali metals since they generally exhibit properties characteristic of covalent and ionic compounds [3]. Thus, organolithium compounds are soluble not only in basic solvents such as ethers, but also in hydrocarbon solvents [3, 4]. They are associated into aggregates in the solid state, in solution, and in the gas phase [5-7]. The structure and the nature of the bonding in organolithium aggregates has been deduced primarily from X-ray crystallographic studies. Both ethyllithium [8] and methyllithium [9] possess approximately tetrameric structures in the solid state with lithium occupying the apices of a tetrahedron with the alkyl groups bonded to the faces of the tetrahedron. In contrast, cyclohexyllithium co-crystallizes with 2 molecules of benzene to form a hexameric structure in the solid state with lithium occupying the apices of an octahedron, six alkyl groups bonded to six faces, and a benzene molecule lying above each of the two vacant faces on opposite sides of the octahedron [10]. Figs. 1 to 3 show these structures.

Because of their oligomeric nature, the bonding in alkyllithium compounds is a subject of considerable interest. The aggregates can be characterized as electron-deficient since there are more nearest-neighbor atom-atom connections than there are valence electron pairs available for bonding [11]. The covalent nature of organolithium bonding is a consequence of the fact that among the alkali metals lithium has the smallest radius, the highest ionization potential, the greatest electronegativity, and the availability of relatively low-lying unoccupied p orbitals for bonding [12]. This unique nature of carbon-lithium bonding is directly responsible for the ability of only lithium among the alkali metals to stereospecifically polymerize dienes to high 1,4 products [13, 14]. Stucky, et al. [10] have described the bonding in alkyllithium aggregates in terms of localized four-centered bonds, involving a triangle of lithium atoms and the bridging carbon atom of an alkyl ligand. It was proposed that the lithium atoms are sp² hybridized. This simplified model provides both a p orbital lobe and an sp² orbital which can interact with Lewis bases.

There is considerable controversy regarding the degree of covalent character in a carbon-lithium bond [15-18]. An uncritical comparison of electronegativities does indicate a high degree of ionic character as do extended Hückel molecular orbital

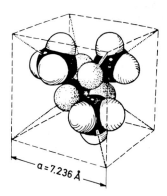

Fig. 1. Model of a tetrameric unit in the crystal structure of methyllithium. (Reprinted with permission from Ref. [9], Copyright 1970, Elsevier Sequoia S.A.)

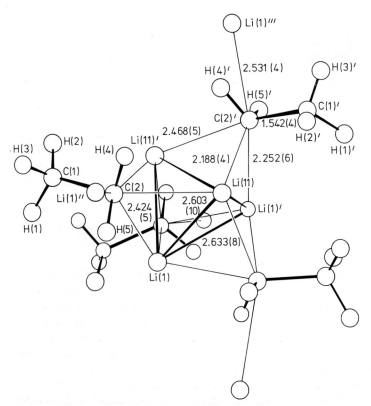

Fig. 2. Crystal structure of ethyllithium (Reprinted with permission from Ref. [8], Copyright 1981 Elsevier, Sequoia S.A.)

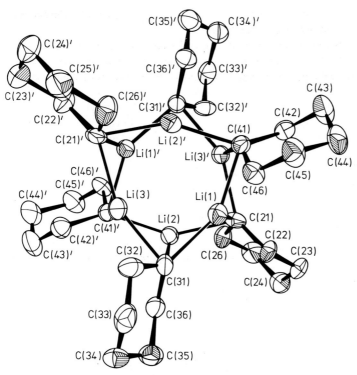

Fig. 3. Crystal structure of cyclohexyllithium. (Reprinted with permission from Ref. [10], Copyright 1974, American Chemical Society)

calculations [19, 20]. However, some of the physical properties, e.g., volatility or solubility in hydrocarbon solvents, are not compatible with highly ionic character. Thus, more complex types of bonding would seem to be present. For example molecular orbital calculations [17–20] have seemingly confirmed that in the aggregated state *covalent* bonding is present with only fractional charges on the carbon and lithium atoms.

The most rigorous *ab initio* calculations indicate a substantial amount of covalent bonding (ca. 40%) in methyllithium, [17] in contrast to reports that this C-Li bond has "essentially no shared-electron covalent character" [15, 16]. The presence of both vacant p orbitals and partial positive charge on lithium contribute to the electron deficiency of lithium. Approximate molecular orbital calculations yield lithium hybridizations of $sp^{1.9}$, $sp^{1.7}$ and sp^1 and carbon hybridizations of $sp^{3.4}$, $sp^{2.9}$ and $sp^{3.7}$ for the dimer, tetramer, and hexamer, respectively [18]. The atomic charges on lithium were calculated to be +0.23, +0.25 and +0.15 electronic units and the atomic charges on carbon were calculated to be −0.27, −0.31 and −0.23 electronic units for the dimer, tetramer, and hexamer, respectively [18]. These calculations offer a possible rationalization for the increased reactivity of less aggregated alkyllithiums (e.g. in initiation of anionic polymerization): the less aggregated species, in general, have a higher degree of ionic character in the carbon-lithium bond. The energetics of dissociation of these aggregates will be discussed in the following section.

Structures analogous to those observed by X-ray crystallography have been proposed for the corresponding associated species in solution; colligative property measurements (see Table 1) have shown that simple, straight-chain, unhindered alkyllithium compounds are associated into hexameric aggregates in hydrocarbon solutions. In contrast, alkyllithium compounds with branching at either the α- or β-carbon tend to associate into tetramers. While s-butyllithium and t-butyllithium are tetrameric in hydrocarbon solution [30,31], both trimethylsilylmethyllithium, $[(CH_3)_3SiCH_2Li]$, and isopropyllithium exhibit [7,26,32,33] equilibria between hexamers and tetramers which are dependent on solvent, concentration and presumably tempera-

Table 1. Association States of n-Alkyl Organolithium Compounds

Compound	Solvent	n^a	Method[b]	Ref.
C_2H_5Li	Benzene	~6	F	[2b]
	Benzene	~6	F	[22]
	Benzene	4.5–6.0	F	[23]
	Benzene	6.0	F	[24]
	Cyclohexane	6.0	F	[25]
	Benzene	6.1	F	[7,26]
	Cyclohexane	6.0	F	[7,26]
$n\text{-}C_4H_9Li$	Benzene	6.3	I	[27]
	Benzene	~7	B	[28]
	Cyclohexane	6.2	I	[29]
	Benzene	6.0	F	[7,26]
$n\text{-}C_6H_{11}Li$	Benzene	6.0	V	[29]
$n\text{-}C_8H_{17}Li$	Benzene	6.0	V	[29]

[a] Average degree of aggregation;
[b] F, freezing point depression; I, isopiestic; B, boiling point elevation; V, vapor pressure depression.

ture. Both trimethylsilymethyllithium and isopropyllithium are tetrameric in benzene solution but exhibit an increase in degree of aggregation (presumably to hexamers) at concentrations higher than 0.06 m or 0.1 m, respectively. Trimethylsilylmethyllithium is hexameric in cyclohexane, while isopropyllithium is tetrameric in cyclohexane up to a concentration of about 0.03 m above which the degree of association increases. It has been reported that 2-methylbutyllithium [34], although hexameric in pentane at 0.89 m (30 °C), has a lower association number (3.2) at lower concentrations (0.048 m) and a higher association number (7.6) at lower temperatures (−12 °C) [34]. Menthyllithium, a sterically hindered species, associates into highly reactive dimers in hydrocarbon solution [35]. Dimeric association is also reported for hydrocarbon solutions of benzyllithium [7], which can be used as a model for poly(styryl)lithium. The degree of association of 3-neopentylallyllithium, a model for poly(dienyl)lithiums, has been reported to be highly concentration dependent in benzene ranging from 3.7 (0.330 m) to 2.14 (0.0495 m) [36]. However, these results are complicated by the fact that the samples employed were contaminated with approximately 10 % t-butyllithium. The dimeric association state has also been observed [37] for 9-(2-hexyl)fluorenyllithium in cyclohexane over the concentration range of 0.01 to 0.1 m.

These results show that the degree of association of organolithium compounds in hydrocarbon solution is quite dependent on the structure of the organic moiety, the solvent, the concentration, and the temperature. The degree of association of alkyl-lithiums can decrease with decreasing concentration, by using a more strongly solvating solvent, by increasing temperature, and either by substituting a more hindered organic group or one that is capable of delocalizing electrons.

The relative reactivities of alkyllithiums as polymerization initiators [38, 39] are intimately linked to their degree of association as shown below with the average degree of association in hydrocarbon solution, where known, indicated in parentheses:

Relative Reactivity of Alkyllithium Initiators

Styrene: menthyllithium(2) > *sec*-BuLi(4) > *i*-PrLi(4–6) >
  i-BuLi > *n*-BuLi(6) > *t*-BuLi(4).

Dienes: menthyllithium(2) > *sec*-BuLi(4) > *i*-PrLi(4–6) >
  t-BuLi(4) > *i*-BuLi > *n*-BuLi(6).

It is clear that, in general, the less associated alkyllithiums are more reactive as initiators than the more highly associated species. The effect of solvent on initiator reactivity is also consistent with the importance of association phenomena. Aromatic solvents, which tend to decrease the average degree of association and promote dissociation processes of aggregates, are reported to lead to initiation rates which are two to three powers of ten faster than in aliphatic solvents [30, 40].

Table 2. Association States of Polymeric Organolithium Active Centers in Hydrocarbon Solvents

Anionic Moiety	Association State	Analytical Method	Ref.
1,3-Butadiene	2 to 4[a]	Cryoscopy	[36]
	2	Viscosity	[41,42]
	4	Light scattering	[43]
	2	Linking reactions	[44]
2-Methyl-1,3-butadiene	2	Viscosity	[45,47,48]
	1.6, 2.4, 3.7	Light scattering	[49]
	~3[b]	Light scattering	[43]
	2	Light scattering	[42]
	2	Light scattering	[50]
	2	Linking reactions	[44]
Styrene	2	Viscosity	[42,45–48]
	2	Light scattering	[43,51]
4-Vinyl biphenyl	2	Viscosity	[48]
1,1-Diphenylethylene	2	Viscosity	[52]
1,3-Pentadiene	2	Viscosity	[48]
2,3-Dimethylbutadiene	2	Viscosity	[48]
2-Methyl, 3-methyl, 4-methyl, and 2,4-dimethyl-1,3-pentadienes	2	Viscosity	[48]
1,3-Cyclohexadiene	~1.7	Viscosity	[48]
2,4-Hexadiene	~1.7	Viscosity	[48]

[a] Extent of aggregation is concentration dependent;
[b] Taken to represent a mixture of dimers and tetramers.

Polymeric organolithium compounds are also associated into aggregates in hydrocarbon solution. The most recent evidence from cryoscopy, light scattering and concentrated solution viscosity measurements, and also from a study of linking reactions,
indicates that both poly(styryl)lithium and poly(dienyl)lithium species are associated
into dimers in hydrocarbon solution (Table 2). A dependence of the degree of
association on concentration for poly(isoprenyl)lithium has been reported [50] from
a combination of small angle light scattering and dilute solution viscosity measurements. Taking into account the expected relative behavior of star branched and linear
polymers, an extrapolated value of 4 for the degree of association at $10^{-2}\ m$ was
obtained. These results parallel the findings of Glaze and co-workers [36] for neopentylallyllithium.

Lewis bases exert dramatic effects on the rate, stereochemistry, and reaction pathway in organolithium chemistry [4]. A partial explanation for these observations can
be deduced from the effects of Lewis bases on the degree of association of
organolithium compounds as shown in Table 3. In general, the presence of basic
molecules tends to decrease the average degree of association of organolithium
compounds. Thus, simple alkyllithiums which are hexameric in hydrocarbon solution

Table 3. Degree of Association of Organolithium Compounds in Polar Solvents

Compound	Concentration[a]	Solvent	Degree of Association	Ref.
2-Methyl-2-phenyl, propyllithium[b]	0.1 M	Diethyl ether	2	[53]
Allyllithium	1.5 M	Diethyl ether	>10	[54]
	1.6 M		>12	[55]
	0.1 M		2.0	[55]
	0.8 M	Tetrahydrofuran	> 1.4	[54]
Benzyllithium	0.1–0.7 M	Tetrahydrofuran	1	[56]
trans-1-Propenyllithium	0.24–0.63 M	Diethyl ether	4	[55]
2-Propenyllithium	0.04 M	Diethyl ether	2	[55]
	0.84 M		4	
Phenyllithium	0.1–0.7 M	Tetrahydrofuran	2	[56]
Ethyllithium	0.38–1.08 M	Diethyl ether	4	[55]
Methyllithium	0.1–1.2 M	Tetrahydrofuran	4	[56]
	0.2–0.8 M	Diethyl ether	4	
n-Butyllithium	0.2–0.6 M	Diethyl ether	4	[56]
9-(2-Hexyl)Fluorenyllithium	0.004 m	Tetrahydrofuran	1.39	[37]
	0.04 m		1.63	
$LiCH_2CN$	0.056–0.186 m	Dimethyl sulfoxide	4	[57]
$C_6H_5CH(Li)CN$		Dimethyl sulfoxide	2	[57]
$(CH_3)_2C(Li)CO_2CH_3$	0.07–0.34 m	Tetrahydrofuran	3.5	[58]
$(CH_3)_2C(Li)CO_2C_2H_5$[c]	0.05–0.31 m	Tetrahydrofuran	3.5	[58]
$(CH_3)_2C(Li)CO_2C(CH_3)_3$	0.05–0.26 m	Tetrahydrofuran	2.3	[58]
$o\text{-}LiC_6H_4CH_2N(CH_3)_2$	0.0248 M	Dimethyl sulfoxide	1.87	[59]
	0.0819 M		4.45	
$o\text{-}LiCH_2C_6H_4CH_2N(CH_3)_2$	0.045 M	Dimethyl sulfoxide	1.84	[59]
$o\text{-}CH_3C_6H_4Li$	0.06–0.72 M	Diethyl ether	2	[60]
$o\text{-}\alpha C_{10}H_8Li$	0.02–0.21 M	Diethyl ether	2	[60]

[a] Expressed in terms of the monomeric (unassociated) species;
[b] Tetrameric at concentrations of ca. 0.07 M in n-pentane;
[c] Exhibits an association state of 6.6 in benzene over a concentration range of 0.1–0.42 m

(Table 1) are converted to solvated tetramers in basic solvents such as ethers. Species which are dimeric in hydrocarbon solution, such as benzyllithium and poly (styryl)lithium, are converted into the unassociated species in tetrahydrofuran solution [42, 56]. The claim [61] that poly(styryl)lithium active centers can exist in an associated form in tetrahydrofuran is known [62] to be incorrect.

Although this review does not deal specifically with the polymerization of polar monomers, it is important to note that many heteroatom lithium derivatives are also associated in solution as shown in Table 4. The strong association of lithium alkoxides, even in polar solvents, is especially noteworthy since lithium alkoxides are product of oxidation (by contamination) of organolithium compounds (see Eq. (88)). Thus, cross-association of lithium alkoxides with alkyllithium initiators or propagating polymeric organolithium would be expected to alter the average degree of association and complicate the corresponding kinetics.

Table 4. Association Behavior of Organolithium Compounds with Non-Carbon Bonded Lithium

Compounds	Solvent	Association State	Ref.
$(CH_3)_3COLi$	Benzene	6	63,64)
	Tetrahydrofuran	4	64)
⬡—NHLi	Cyclohexylamine	2.6 to 3	65)
$[(CH_3)_3Si]_2NLi$	Benzene	2	66)
	Diethyl ether	1.6	

3 Enthalpies of Dissociation of Organolithium Aggregates

A number of theoretical studies of the bonding of alkyllithiums have provided an estimate of the dissociation energies of alkyllithium aggregates in the absence of solvent or Lewis base interactions, i.e., for isolated gas-phase species. The results of these calculations for the energetics of dissociation of aggregates are listed in Table 5. It is of interest to note that the most reliable calculations with optimized geometries predict, for methyllithium, that the process of dissociation of a dimer to two unassociated methyllithiums corresponds to a binding energy of approximately 20 kcal/mole per alkyllithium molecule. Complete dissociation of a tetramer to four unassociated species would require approximately 108–124 kcal/mole, i.e., a binding energy per alkyllithium of 27–31 kcal/mole. The most recent calculations with optimized geometries also reveal that the tetrahedral (T_d) tetrameric and octahedral (D_{3d}) hexameric structures are very close in energy to the corresponding planar structures. The significance of this result is that although complete dissociation of a tetramer (i.e., Eq. (1)) would require about

$$(CH_3Li)_4 \rightarrow 4CH_3Li \tag{1}$$

108 kcal/mole of tetramer (a value, as noted by Brown [70] in 1966, which would prohibit the formation of a significant concentration of the unaggregated species)

Table 5. Energetics of Dissociation of Alkyllithium Aggregates

Organolithium	Dissociation Process	Dissociation Energy (kcal mol^{-1})	Calculation Method	Ref.
Methyllithium	dimer → 2 monomers	39.9	*ab initio* (STO-3G)	[67]
	dimer → 2 monomers	41.6	*ab initio* (4-31G)	[67]
	tetramer → 4 monomers	124.4	*ab initio* (STO-3G)	[67]
	dimer → 2 monomers	37.2	*ab initio* (STO-3G)	[18]
	dimer → 2 monomers	36.0	PRDDO	[18]
	dimer → 2 monomers	39.6	*ab initio* (double zeta basis)	[18]
	tetramer → 4 monomers	108.4	PRDDO	[18]
	dimer → 2 monomers	34.9	*ab initio* (STO-3G)	[68]
	tetramer → 4 monomers	114	*ab initio* (minimum basis)	[69]
Ethyllithium	dimer → 2 monomers	34.4	PRDDO	[18]

dissociation of a tetramer to give one methyllithium molecule and the cyclic trimer (D_{3h}) (Eq. (2)) would only

$$(CH_3Li)_4 \rightarrow (CH_3Li)_3 + CH_3Li \tag{2}$$

require 29 kcal/mole. The energy of dissociation of a tetramer to two dimers (Eq. (3))

$$(CH_3Li)_4 \rightarrow 2(CH_3Li)_2 \tag{3}$$

was calculated to be *ca.* 36 kcal/mole. One of the important conclusions from these theoretical calculations is that the dissociation energies for processes such as those shown in Eqs. (2) and (3) are accessible at normal polymerization temperatures, i.e., one need not invoke complete dissociation processes such as that shown in Eq. (1). It is interesting to note that Brown and co-workers have reported dissociation energies for tetramer-dimer equilibria of 11 and 24 kcal/mole for methyllithium in ether [71] and *t*-butyllithium in cyclopentane [72], respectively. In fact, however, these numbers refer to the processes involved in equilibration of alkyllithiums and could involve a dissociation such as that shown in Eq. (2) for the rate determining step. Another significant result is that planar, cyclic structures may be involved as intermediates in various exchange processes [18]. An additional indication that organolithium species possess high values for their dissociation energies is given by results garnered from mass spectroscopy studies [63, 72, 73]. Ethyllithium, *t*-butyllithium and trimethylsilyl-methyllithium have been found to retain the hexameric or tetrameric state in the gas phase. There was no indication that lesser associated structures, e.g., dimers, or unassociated species exist under the conditions of measurement.

There are relatively few theoretical studies available relating to the dissociation energies of allyllithium species. Theoretical calculations of the CNDO/2 type, which are recognized to give rather inaccurate values of electronic energies, provide a value [74, 75] of 201 kcal/mole for the dissociation energy of *cis*-crotyllithium (Eq. (4))

$$(cis\text{-}C_4H_7Li)_2 \rightleftharpoons 2 \; cis\text{-}C_4H_9Li \tag{4}$$

There is very little experimental evidence relating to the energetics of dissociation of poly(dienyl)lithium species. From the temperature dependence of the flow times of the concentrated solution viscosities of hexane solutions of poly(isoprenyl)lithium, Morton and Fetters [47] reported an estimate of 37 kcal/mole for the dissociation of dimers (Eq. (5)).

$$(PILi)_2 \rightleftharpoons 2 \, PILi \qquad\qquad\qquad (5)$$

Roovers and Bywater [76] examined the temperature dependence of the electronic spectrum of poly(isoprenyl)lithium and were able to calculate an equilibrium constant for the dissociation event. On the basis that the process involved was tetramers $\rightleftharpoons$ dimers, the dissociation enthalpy was determined to be 12.3 kcal/mole in n-octane while a value of 9.0 kcal/mole was found in benzene solution. The latter value was thought to be due to weak solvations of the active centers by benzene. The approach used by Roovers and Bywater [76] is predicated on the assumption that the 272 and 320 nm absorptions represent species differing in their association state.

If the process measured by Roovers and Bywater is reanalyzed on the basis of a monomer-dimer-dissociation equilibrium, their results yield a value of about 11 kcal/ mole. Szwarc [76a-78] has presented, without citing or providing either theoretical or experimental evidence, various values (12, 14–15, and 15–16 kcal/mole) for this step. Meier, using the approach involving the temperature dependence of concentrated solution viscosities, reported [79] a value of 21.8 kcal/mole for the dissociation enthalpy of the poly(styryl)lithium dimers. These combined results will be discussed and compared with direct calorimetric results in a later section of this review.

4 Solvation of Organolithium Compounds

In the alkyllithium initiated polymerizations of vinyl monomers, Lewis bases such as ethers and amines alter the kinetics, stereochemistry, and monomer reactivity ratios for copolymerization. In general, the magnitude of these effects has been directly or indirectly attributed to the extent or nature of the interaction of the Lewis base with the organolithium initiator or with the organolithium chain end of the growing polymer. Unfortunately, all of these observed effects are kinetic in nature, and therefore the observed effects of solvent represent a composite effect on the transition-state *versus* the ground state as shown below in Eq. (6), where δ represents the differential

$$\delta\Delta G^{\ddagger} = \delta\Delta G^{T.S.} - \delta\Delta G^{G.S.} \qquad\qquad\qquad (6)$$

effects of solvent variation and $\Delta G^{T.S.}$, $\Delta G^{G.S.}$ and $\Delta G^{\ddagger}$ represent the free energies of the transition state, of the ground state, and of activation, respectively. Consequently, an increase in rate could be associated with either a decrease in the relative free energy of the transition state, or with an increase in the free energy of the ground state.

Several years ago [80-83], a systematic investigation of the energetics of interaction of Lewis bases with organolithium compounds was undertaken. The enthalpies of interaction of Lewis bases with organolithium compounds as a function of R ([base]/[lithium atoms], Eq. (7)) were measured using high-dilution, solution calorimetry. These measurements provide direct, thermodynamic information regarding the nature of the coordination of bases with organolithium compounds which corresponds to ground-state solvation effects ($\delta\Delta H^{G.S.}$).

$$\text{RLi} + \text{B} \xrightarrow{\Delta H} \text{RLi} \cdot \text{B} \tag{7}$$

The solution calorimeter and operating procedures used for these studies were essentially the same as those described by Arnett, et al. [84]. For investigation of organolithium compounds, the calorimetric vessel was placed inside a recirculating, argon atmosphere glovebox with a gas purification train patterned after the system developed by Brown and coworkers [85]. The calorimeter was connected to the other calorimetric components outside the glovebox via an electrical feedthru. Solutions of organolithiums were analyzed for the amount of carbon-bound lithium using the double titration procedure of Gilman and Cartledge [86] with 1,2-dibromoethane before and after calorimetric runs to check purity and to show that decomposition reactions with the bases were not occurring in the calorimeter. The preparation, purification and calorimetric procedures for Lewis bases, alkyllithiums, and polymeric organolithiums have been described earlier in detail [80−83]. *Without exception* the results of double titration analysis of the calorimetric solutions indicated that within the error limitations of this analytical method no decomposition of organolithium reagents occurs under these conditions.

The reliability of the calorimetric results for these reactive organometallic compounds was further substantiated by the observations that the results were all quite reproducible (± 0.1 kcal/mole) and that the results obtained do not depend on (1) the source or method of purification of the base or the solvent; (2) the source or method of purification of the alkyllithium; and (3) the sodium content of the lithium metal used to prepare the alkyllithiums. Furthermore, the calorimetric equipment was regularly calibrated with internationally accepted standards for calorimetry.

The initial enthalpies of interaction of small amounts (< 1 mmol) of Lewis bases with dilute solutions (0.04–0.08 M) of hexameric and tetrameric alkyllithiums in hydrocarbon solution at 25 °C are shown in Table 6. For the initial enthalpies listed in Table 6, the ratio of the concentration of base to the concentration of lithium atoms in solution (R) is less than 0.08. These enthalpies provide a quantitative measure of the relative strength of these interactions. *The basicity order which emerges from the relative initial enthalpies of interaction was found to be the same for every alkyllithium compound examined,* viz. tetrahydrofuran (THF) > 2-methyltetrahydrofuran [(2-CH$_3$THF] > 2,5-dimethyltetrahydrofuran [2,5-(CH$_3$)$_2$THF] > diethyl ether > triethylphosphine > triethylamine > tetrahydrothiophene [82]. This result was unexpected since the alkyllithium compounds examined included species, some of which are present as hexameric aggregates, and some of which exist predominantly as tetrameric aggregates in hydrocarbon solution. The lack of dependence on structure or degree of association suggested that this basicity order represented a fundamental property of alkyllithium-base interactions. It was therefore of interest to determine if this basicity order would be applicable to the interaction of polymeric organolithiums with bases, and especially if the order phosphorus > nitrogen would be maintained. This question will be discussed later.

Table 6. Enthalpies of Interaction of Bases with Alkyllithiums at low R Values [80–82]

Alkyllithium	$-\Delta H$ (kcal/mole)[a,b]						
	THF	2-CH$_3$-THF	2,5-(CH$_3$)$_2$-THF	Et$_3$P	Et$_3$N	(CH$_2$)$_4$S	Et$_2$O
C$_2$H$_5$Li[e]	7.4[c](0.08)	—	—	1.4[d](0.04)	1.1[d](0.05)	—	—
n-C$_4$H$_9$Li[f]	7.5[c](0.08) 7.6[d](0.07)	6.3[c](0.06)	4.6[c](0.05)	1.5[d](0.04)	1.2[d] (0.04)	0.4[d](0.07)	1.8[d](0.06)
(CH$_3$)$_3$SiCH$_2$Li[e]	10.3[c](0.08)	10.4[c](0.06)	9.7[c](0.05)	4.6[c](0.05)	3.5[c](0.05)	1.6[c](0.08)	6.3[c](0.06)
(CH$_3$)$_3$SiCH$_2$Li[g]	9.4[c](0.08)	9.6[c](0.08)	8.7[c](0.05)	3.3[c](0.05)	2.1[c](0.05)	0.9[c](0.08)	5.3[c](0.06)
(CH$_3$)$_2$CHLi[g]	8.5[c](0.08)	—	6.1[c](0.05)	—	—	—	—
(CH$_3$)$_3$CLi[e]	0.1[d]	—	—	0.2[d]	0.0[d]	—	—

[a] All enthalpies obtained at 25° by addition of 0.050 ml or less into 195 ml of 0.04 M or 0.08 M alkyllithium (base/Li atom ratio $\leqq$ 0.08) [b] The numbers in parenthesis represent the [Base]/[Li atom] ratio for each measurement; [c] 0.04 M; [d] 0.08 M; [e] Cyclohexane solvent; [f] Hexane solvent; [g] Benzene solvent

One point of particular interest was the fact that the enthalpies of interaction of tetrameric organolithiums were more exothermic than the corresponding enthalpies of interaction for hexameric species. The one exception was t-butyllithium which did not interact significantly with any of the bases. These results suggest that tetrameric organolithiums interact more strongly than hexameric species with a given base. However, this conclusion is tempered by the fact that calorimetry itself cannot define the process involved in base coordination. What is known about the effect of bases on association of alkyllithiums is that: (a) alkyllithiums which are hexameric in hydrocarbon solution exist as tetrameric aggregates in basic solvents such as monodentate ethers and amines; (b) hexameric alkyllithiums in hydrocarbon solution are converted into tetramers at R values $\leqq 1.0$; and (c) there is no evidence for degrees of association less than four for simple alkyllithiums in monodentate ethers and amines [26]. It can therefore be concluded that tetrameric alkyllithiums in hydrocarbon solution interact with bases to form solvated tetramers (Eq. (8)).

$$(RLi)_4 + B \rightarrow (RLi)_4 \cdot B \tag{8}$$

However, for hexameric alkyllithiums in hydrocarbon solution, bases can coordinate to form either a solvated hexamer (Eq. (9));

$$(RLi)_6 + B \rightarrow (RLi)_6 \cdot B \tag{9}$$

or a solvated tetramer (Eq. (10));

$$2/3\,(RLi)_6 + B \rightarrow (RLi)_4 \cdot B \tag{10}$$

From studies of the sensitivity of the coordination process to the steric requirements of the base (based on comparison of enthalpies for tetrahydrofuran *versus* 2,5-dimethyltetrahydrofuran), it has been concluded that interaction of bases with hexameric alkyllithiums (e.g., n-butyllithium) at low R values involves coordination with the intact hexamer (Eq. (9)) [82]. Since hexamers are likely to be more sterically hindered than tetramers, these results offer a possible explanation for the observation that, in general, less aggregated organolithiums are more reactive initiators than more associated species (e.g., dimeric menthyllithium $>$ tetrameric sec-butyllithium $>$ hexameric n-butyllithium). The calorimetric studies of the base interactions with alkyllithiums indicate that less aggregated organolithiums *appear* to be less hindered and interact more exothermically with bases. It is probable, therefore, that less aggregated alkyllithiums also interact more strongly with monomers. Thus, *initiator reactivity may be explained partially in terms of the steric requirements of the aggregated organolithium*, which in turn affect the strength of the interaction with incoming monomer.

It is well known that bidentate bases, such as N,N,N',N'-tetramethylethylenediamine (TMEDA), are very effective coordinating bases for organolithium compounds and that they generally greatly enhance the reactivity of organolithiums [87]. Calorimetric studies of the enthalpies of interaction of TMEDA and diglyme (1,2-diethoxyethane) have provided quantitative evidence for the strong coordination of these bases with organolithiums as shown in Table 7 [80-82]. Diglyme interacts 4.7 kcal/mole more exothermically than diethyl ether with n-butyllithium, while TMEDA interacts 9.1 kcal/mole more exothermically than triethylamine. Detailed interpretation of these

Table 7. Comparison of the Initial Enthalpies of Interaction of Bidentate versus Monodentate Bases with n-Butyllithium[a]

Base	ΔH (kcal/mole)[b]
Tetrahydrofuran	7.6 (0.07)
Diglyme	6.5 (0.06)
Diethyl ether	1.8 (0.06)
Triethylamine	1.2 (0.04)
TMEDA	10.3 (0.04)

[a] 0.08 M n-C$_4$H$_9$Li in n-hexane;
[b] The numbers in parenthesis represent the [Base]/[Li atom] ratio

results is limited by the lack of consistent data for the effect of these bases on the degree of association of alkyllithiums.

The most dramatic effects of Lewis bases in organolithium chemistry are observed in polymerization reactions. Aside from colligative property measurements, there is little direct quantitative information on the nature of the organolithium-base interactions responsible for the observed effects. The calorimetric method has been used also to examine the fundamental nature of the interaction of bases with polymeric organolithium compounds [83, 88, 89]. Information is now available on the ground-state interaction of bases with poly(styryl)lithium (PSLi), poly(isoprenyl)lithium (PILi) and poly(butadienyl)lithium (PBDLi).

Initial enthalpies for addition of small amounts of bases to dilute solutions (0.2 M) of polymeric organolithiums at low R values ([base]/[Li]) provide direct information on the strength of the base interactions as well as the steric requirements of the bases. Data for initial enthalpies of interaction for a variety of bases with poly(styryl)lithium in benzene are listed in Table 8 [88, 89]. It is especially significant to note that the basicity order observed for poly(styryl)lithium (TMEDA > diglyme > THF > 2,5-Me$_2$THF > dioxane > TMEDP > Et$_3$P > Et$_2$O $\cong$ Et$_3$N) is very similar to the order for simple alkyllithiums; (see Tables 6 and 7) TMEDA > THF >

Table 8. Enthalpies of Interaction of Bases with Poly(styryl)lithium[a]

Base	$(-\Delta H^b/\text{mole})$
N,N,N',N'-tetramethylethylenediamine (TMEDA)	13.3
Dimethoxyethane	9.8
Tetrahydrofuran	4.5
2,5-dimethyltetrahydrofuran	2.3
Dioxane	1.5
Tetramethylethylenediphosphine (TMEDP)	0.9
Triethylphosphine	0.4
Diethyl ether	0.3
Triethylamine	0.3

[a] $\bar{M}_n \sim 4 \times 10^3$;
[b] All enthalpies obtained at 25 °C by addition of 100 μl (ca. 1 mmole) of base into 200 ml of 0.03 M PSLi([Base]/[Li] ca. 0.2)

diglyme $>$ 2,5-Me$_2$THF $>$ Et$_2$O $>$ Et$_3$P $>$ Et$_3$N). The notable exceptions are di-glyme and diethyl ether. Diglyme can apparently interact more effectively relative to tetrahydrofuran with this less associated, and therefore less hindered, polymeric organolithium. Triethylphosphine, diethyl ether, and triethylamine all interact very weakly with poly(styryl)lithium; the enthalpies for these bases are all less exothermic than the corresponding enthalpies for alkyllithiums. It is interesting to note that although triethylphosphine interacts more exothermically than triethylamine with both simple and polymeric organolithiums, TMEDA interacts much more exother-mically than the corresponding diphosphine (TMEDP). The uniqueness of bidentate nitrogen bases is also apparent from the relative basicity order observed for bidentate bases interacting with polymeric organolithiums: TMEDA $>$ diglyme $\gg$ dioxane $>$ diphosphine. In order to rigorously interpret these calorimetric results, it is essential to know what the nature of the base-PLi adduct is, i.e., the state of association and the stoichiometry. These factors will be considered in the following sections.

Comparison of the enthalpies of interaction of 2,5-dimethyltetrahydrofuran vs. tetrahydrofuran provides a useful technique for determining the relative steric re-quirements of base coordination with a variety of organolithium compounds [82]. This method rests on the assumption that the coordination process, whatever its nature, is the same for 2,5-dimethyltetrahydrofuran as it is for tetrahydrofuran with a given organolithium compound. With this assumption, deductions can be made about the coordination process itself by comparison of results for organolithium compounds with different steric requirements and different degrees of association. The differences in enthalpies of interaction for polymeric organolithiums are shown in Table 9. The results for poly(styryl)lithium and poly(isoprenyl)lithium have been reported previously [88]. For poly(styryl)lithium, it is well established that the degree of associa-tion is two in hydrocarbon solution and one in the presence of excess tetra-hydrofuran [42]. Furthermore, concentrated solution viscosity measurements have shown that the equilibrium constant (K) for the process shown in Eq. (11)

$$(\text{PSLi})_2 + 2\,\text{THF} \rightarrow 2(\text{PSLi} \cdot \text{THF}) \tag{11}$$

has a value [42] of approximately 160 LM^{-1}. This equilibrium constant is such that essentially all of the tetrahydrofuran will be converted to the adduct PSLi · THF under our calorimetric conditions (1.2 mmoles of THF added to 200 ml of a 0.03 M solution of PSLi). Therefore, the process involved in the interaction of dimeric

Table 9. $\Delta\Delta$H of Interaction of Polymeric Organolithiums

PLi	$\Delta\Delta$H (kcal/mole)[a]
PSLi	2.2
PILi	3.2
PBDLi	1.5

[a] $\Delta\Delta$H represents the difference in enthal-pies of interaction of PLI with THF vs. 2,5-Me$_2$THF

poly(styryl)lithium with tetrahydrofuran, and presumably 2,5-dimethyltetrahydrofuran, involves conversion into solvated, monomeric poly(styryl)lithium (Eq. (11)). The relatively large difference in enthalpies between tetrahydrofuran and 2,5-dimethyltetrahydrofuran (2.2 kcal/mole) indicates that the base coordination process shown in Eq. (11) for poly(styryl)lithium is quite sensitive to the steric requirements of the base.

The degree of association of poly(dienyl)lithium compounds in hydrocarbon solutions is a matter of current controversy (Table 2). Average association states of both two and four have been reported based on light scattering and concentrated solution viscosity measurements. An average degree of association of two for poly-(dienyl)lithiums has emerged from comprehensive and self-consistent studies utilizing combinations of endcapping and linking techniques coupled with concentrated solution viscosity measurements [44].

The effect of tetrahydrofuran on the extent of association of poly(isoprenyl)lithium in n-hexane has been determined by concentrated solution viscosity measurements. The equilibrium constant for the interaction of THF with poly(isoprenyl) lithium as shown in Eq. (12)

$$(\text{PILi})_2 + 2\,\text{THF} \rightarrow 2(\text{PILi} \cdot \text{THF}) \tag{12}$$

has a value [47] of approximately 0.5 LM^{-1}. This relatively small equilibrium constant suggests that complexation by tetrahydrofuran is occurring primarily with the intact dimer (Eq. (13)) rather than with the unassociated species (Eq. (12)) as observed for poly(styryl)lithium (Eq. (11)).

$$(\text{PILi})_2 + \text{THF} \rightarrow (\text{PILi})_2 \cdot \text{THF} \tag{13}$$

This conclusion provides an explanation for the calorimetric observation that base coordination of poly(isoprenyl)lithium is more sensitive to the steric requirements of the base ($\Delta\Delta H = 3.2$ kcal/mole) than is the coordination process for poly(styryl) lithium ($\Delta\Delta H = 2.2$ kcal/mole), since monomeric poly(isopropenyl) lithium would be expected to be less hindered than unassociated poly(styryl)lithium. However, dimeric poly(isoprenyl)lithium could very well be more hindered toward base coordination (Eq. (13)) than monomeric poly(styryl)lithium (Eq. (11)).

The nature of the process involved in the interaction of tetrahydrofurans with poly(butadienyl)lithium has been less well characterized, although it can be assumed to be analogous to the process involved with poly(isoprenyl)lithium (Eq. (13)). If this is correct, then the interaction of tetrahydrofuran with poly(butadienyl)lithium can be described in terms of Eq. (14).

$$(\text{PBDLi})_2 + \text{THF} \rightarrow (\text{PBDLi})_2 \cdot \text{THF} \tag{14}$$

The decreased steric requirements for this base coordination process ($\Delta\Delta H = 2.1$ kcal/mole) compared to the analogous interaction for poly(isoprenyl)lithium ($\Delta\Delta H = 3.2$ kcal/mole) are consistent with a poly(butadienyl)lithium chain end being less sterically demanding than a poly(isoprenyl)lithium chain end. Several factors can be considered in favor of the same degree of association for the base adduct for poly(butadienyl)lithium and poly(isoprenyl)lithium. First, increasing steric requirements of the organic moiety in an organolithium compound generally lead to decreased degrees of association. Therefore, if their degrees of association are different, poly(isoprenyl)lithium would be expected to be less associated than poly(butadienyl)lithium. If, however, poly(butadienyl)lithium interacts with THF to form an adduct which is more associated than the corresponding adduct for poly(isoprenyl)lithium, one would expect the steric requirements to be greater for base coordination to form this more highly associated adduct. Since the steric requirements for coordination with poly(butadienyl)lithium are less than they are for poly(isoprenyl)lithium, it is reasonable to conclude that the same type of coordination process is involved (e.g. Eqs. (13) and (14)). It should be noted, however, that this discussion cannot rule out the process shown in Eq. (15).

$$1/2(PBDLi)_4 + THF \rightarrow (PBDLi)_2 \cdot THF \tag{15}$$

i.e., a higher average degree of association for poly(butadienyl)lithium *versus* poly(isoprenyl)lithium but formation of the same type of base adduct (Eq. (15) *versus* Eq. (13)).

N,N',N'-Tetramethylethylenediamine and other analogous bidentate nitrogen compounds are unique among Lewis bases in their ability to promote reactions of organolithiums[87]. As described later, there are many conflicting reports regarding the nature of the interaction of TMEDA with organolithium compounds. One interesting facet of this chemistry is the stoichiometric dependence of the effect of TMEDA on the kinetics and microstructure of organolithium-initiated polymerizations of styrene and diene monomers. Schué and coworkers[90] have shown that the effects of TMEDA on rates, microstructure and nuclear magnetic resonance (NMR) and ultraviolet (UV) spectra for isoprene polymerization are dependent on the base to lithium atom ratio R ([TMEDA]/[Li]) with maxima or minima at $R = 0.5$. Similar results and unusual temperature effects were observed by Antkowiak et al.[91] for butadiene polymerizations in the presence of TMEDA-organolithium complexes. In contrast, Helary and Fontanille[92] have reported that TMEDA can increase or decrease the rate of polymerization of poly(styryl)lithium depending on the organolithium concentration. The rate maximum or minimum occurred at $R = 1.0$, not at $R = 0.5$ as observed for poly(isoprenyl)lithium. These results suggested that the interactions of TMEDA with polymeric organolithiums may have specific stoichiometric dependencies which could be determined using calorimetry.

The calorimetric data for the exothermic enthalpies of interaction of poly(styryl)lithium, poly(isoprenyl)lithium and poly(butadienyl)lithium with TMEDA as a function of R are shown in Figs. 4, 5 and 6, respectively. It is obvious from these data that the stoichiometric dependence of the interaction of TMEDA with these polymeric organolithiums is different for each of these species. In order to try to interpret the significance of these results, it is necessary to again consider the question of the state of association of these polymeric organolithiums. As discussed earlier, poly(styryl)lithium is predominantly associated into dimers in benzene solution[42, 43, 45−48, 51]. In the presence of an excess of the moderately strong base tetrahydrofuran, the self-association of poly(styryl)lithium is eliminated[42]. In view of the stronger coordinating ability of TMEDA, therefore, it would be expected that the following process is involved in the interaction of TMEDA with poly(styryl)lithium, (Eq. (16)).

$$(PSLi)_2 + 2\ TMEDA \rightarrow 2(PSLi \cdot TMEDA) \qquad (16)$$

This is consistent with the relative ease of dissociation of poly(styryl)lithium in the presence of bases and also with the concentration dependence of enthalpy versus R plot (Fig. 4) with a break observed at an R value of $ca.$ 1.0. These calorimetric results are also in agreement with the stoichiometric dependencies observed by Helary and Fontanille [92] from their kinetic and spectroscopic studies. For example, they reported that the UV wavelength of maximum absorption of poly(styryl)lithium shifted upon additions of TMEDA until an R value of 1.0 and then was constant. They also observed that TMEDA additions increased or decreased the rate of polymerization, depending on [PSLi]; the increase or decrease in the rate leveled off at an R value of $ca.$ 1.0. All

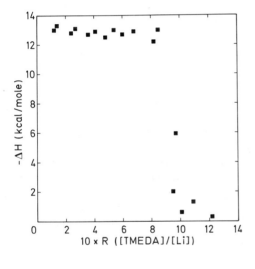

Fig. 4. Enthalpies of interaction of TMEDA as a function of R([TMEDA]/[Li]) for 0.02 M benzene solutions of poly(styryl)lithium

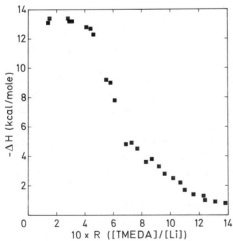

Fig. 5. Enthalpies of interaction of TMEDA as a function of R([TMEDA]/[Li]) for 0.02M benzene solutions of poly(isoprenyl)lithium

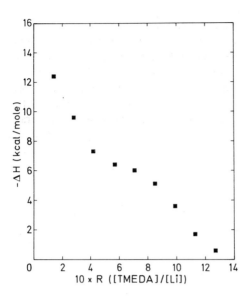

Fig. 6. Enthalpies of interaction of TMEDA as a function of R([TMEDA]/[Li]) for 0.02M benzene solutions of poly(butadienyl)lithium

of these results can be interpreted in terms of the process shown in Eq. (16), i.e., they are directly related to the stoichiometry of the ground-state interaction of TMEDA with poly(styryl)lithium as determined by calorimetry.

As discussed previously, there is disagreement in the literature regarding the average degree of association of poly(dienyl)lithium compounds in hydrocarbon solution. Aggregation states of both two and four have been reported for poly(isoprenyl)lithium in hydrocarbon solution (Table 2). The stoichiometry inferred for the enthalpy of interaction of TMEDA with poly(isoprenyl)lithium appears to be different than the corresponding process for poly(styryl)lithium. The concentration dependence of the enthalpy versus R plot (Fig. 5) exhibits a break at an R value of *ca.* 0.5 for poly(isoprenyl)lithium. Based on comparison of the steric requirements of base coordination as determined by calorimetry, it was concluded that tetrahydrofuran may interact with poly(isoprenyl)lithium to form a THF-solvated aggregate, presumably the dimer (Eq. (13)). It is possible that a similar type of coordination process is involved with TMEDA. Obviously calorimetry itself cannot identify the coordination process; however, any explanation describing the interaction of TMEDA must be consistent with the calorimetric evidence (break at R = 0.5, Fig. 5). Several schemes can be used to describe the nature of the interaction of TMEDA with poly(isoprenyl)-lithium. The major variables in these different schemes are the degrees of association of uncoordinated and TMEDA-coordinated poly(isoprenyl)lithium (Eq. (17)–(20)):

$$2\,\text{TMEDA} + (\text{PILi})_4 \rightarrow 2[(\text{PILi})_2 \cdot \text{TMEDA}] \tag{17}$$

$$2\,\text{TMEDA} + (\text{PILi})_4 \rightarrow [(\text{PILi})_4 \cdot 2\,\text{TMEDA}] \tag{18}$$

$$\text{TMEDA} + (\text{PILi})_2 \rightarrow [(\text{PILi})_2 \cdot \text{TMEDA}] \tag{19}$$

$$2\,\text{TMEDA} + (\text{PILi})_2 \rightarrow 2(\text{PILi} \cdot \text{TMEDA}) \tag{20}$$

It has been observed that the concentrated solution viscosity decreases upon addition of TMEDA to solutions of poly(isoprenyl)lithium [93]. This would be consistent with the process shown in Eq. (17) or (20) and not with Eqs. (18) or (19). The decrease in viscosity would be consistent with interaction of TMEDA to form an unassociated complex (Eq. (20)), but this does not seem to be in accord with the stoichiometry observed by calorimetry. It is noteworthy that the break observed by calorimetry at R = 0.5 is consistent with the stoichiometric dependence of spectral, kinetic and microstructure effects [90]. Again this shows that these kinetic effects are related to the stoichiometry of formation of base-organolithium adduct, i.e. that they are ground-state solvation effects.

The situation with respect to the association behavior of poly(butadienyl)lithium is analogous to that of poly(isoprenyl)lithium. Therefore, the same ambiguity applies to the nature of the interaction of poly(butadienyl)lithium with TMEDA. The concentration dependence of the enthalpy versus R plot (Fig. 6) is very complex. More insight into the nature of this comparison can be obtained from enthalpimetric titration plots of cumulative heat vs. R value for these polymeric organolithiums (Fig. 7). Poly(styryl)lithium exhibits relatively simple behavior in this treatment also, since here again as in Fig. 4 a break is observed at an R value of ca. 0.9–1.0. Similarly, poly(isoprenyl)lithium exhibits a break at an R value of ca. 0.5–0.6 as observed in Fig. 5. The enthalpimetric titration data for poly(butadienyl)lithium is still somewhat ambiguous. In one respect the curve for poly(butadienyl)lithium is similar to that of poly(isoprenyl)lithium. The initial and final segments of the curve intersect at an R value of ca. 0.5–0.6, corresponding to processes such as those shown in Eqs. (17)–(19) for poly(isoprenyl)lithium. However, this curve could also be interpreted as possessing two breaks; one at an R value of ca. 0.25–.3 and another at an R value of ca. 0.9–1.0, suggesting perhaps either a greater association multiplicity or more strongly

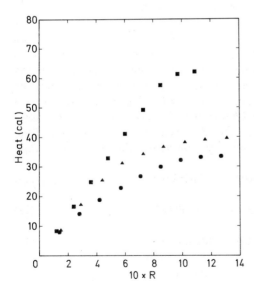

Fig. 7. Enthalpimetric titration plot for interaction of TMEDA with 0.02M solutions of poly(styryl)lithium (■), poly(isoprenyl)-lithium (▲), and poly(butadienyl)lithium (●)

associated state for poly(butadienyl)lithium *versus* poly(isoprenyl)lithium. At the present time it is not possible to say which interpretation is correct.

If all of these polymeric organolithiums are converted into unassociated, TMEDA-coordinated species at an R value of *ca.* 1.0, then the data in Fig. 7 can be used to estimate the *difference* in enthalpies of association for these aggregates. At an R value of 1.0, the cumulative heat in calories can be associated with the process shown in Eq. (21), where

$$(PLi)_n + n(TMEDA) \rightarrow n[PLi \cdot TMEDA] \tag{21}$$

PLi represents any polymeric organolithium. For 4 mmoles of organolithium these cumulative heats are 61.6 calories, 37.6 calories, and 32.4 calories for PSLi, PILi, and PBDLi, respectively. Therefore, the total heats involved correspond to 15.4 kcal/mole, 9.4 kcal/mole, and 8.1 kcal/mole for PSLi, PILi, and PBDLi, respectively. Thus, the conversion shown in Eq. (21) is 6.0 kcal/mole more exothermic for PSLi *versus* PILi, and 7.3 kcal/mole more exothermic for PSLi *versus* PBDLi. The process shown in Eq. (21) can be dissected conceptually into two thermodynamic steps (Eqs. (22) and (23)),

$$1/n \, (PLi)_n \xrightarrow{\Delta H_d} PLi \tag{22}$$

$$PLi + TMEDA \xrightarrow{\Delta H_c} PLi \cdot TMEDA \tag{23}$$

with corresponding enthalpies ΔH_d and ΔH_c, respectively. If it is assumed that ΔH_c is approximately independent of the nature of the polymeric lithium compound, then the enthalpy *differences* observed can be ascribed to differences in enthalpies of dissociation of the polymeric organolithium aggregates. Using these assumptions, the calorimetric results suggest that the enthalpy of association of PSLi is *at least* 6–7 kcal/mole less exothermic than the corresponding association enthalpies for poly(dienyl) lithiums per mole of PLi. These results can be compared with the work of Meier [79] and Morton and Fetters [47] who determined the enthalpies of dissociation of poly(styryl)lithium (21.8 kcal/mole) and poly(isoprenyl)lithium (36.9 kcal/mole) using concentrated solution viscosity measurements. Their difference in enthalpy of association for poly(isoprenyl)lithium *versus* poly(styryl)lithium (15 kcal/mole) compares very favorably with the corresponding difference determined directly (with attendant assumptions) by calorimetry (12 kcal/mole). It appears, therefore, that *the difference in the enthalpy of association* (exothermic) for poly(styryl)lithium *versus* poly(dienyl)lithiums is in the range of 12–15 kcal/mole. Such a large difference could well account for the differences observed for poly(styryl)lithium *versus* poly(dienyl)lithiums in their interaction with bases as determined by high-dilution, solution calorimetry. These findings are inconsistent with the divergent values of 12, 14–15, and 15–16 kcal/mole advanced by Szwarc [76a–78], for the poly(isoprenyl)-lithium dissociation enthalpy since adoption of his Procrustean assessments would require that the corresponding parameter for poly(styryl)lithium assume the unrealistically low values of 0 to 4 kcal/mole of aggregated species.

5 Initiation Reactions Involving Alkyllithiums

The initiation events involving dienes and styrene in hydrocarbon solvents have been thoroughly and accurately studied by the application of UV and visible spectroscopy. The archetype of such studies is the now classic 1960 study of Worsfold and Bywater [94] on the n-butyllithium-styrene system in benzene. The reaction was found to follow the relationship:

$$R_i \propto [n\text{-}C_4H_9Li]_0^{1/6} [M_0] \tag{24}$$

The fractional dependency of the initiation process on the total concentration of initiator was rationalized on the basis of the hexameric association state of n-butyllithium (Table 1) and the following equilibrium:

$$(n\text{-}C_4H_9Li)_6 \rightleftharpoons 6\, n\text{-}C_4H_9Li \tag{25}$$

where it was assumed that only the unassociated form of the organolithium was reactive toward styrene. An additional assumption was that the equilibrium constant for the above dissociation process was no larger than 10^{-6}. However, there does not exist any direct experimental evidence which demonstrates the existence of monomeric n-butyllithium as a distinct molecular species in hydrocarbon solvents nor is there corresponding evidence that n-butyllithium exclusively retains its hexameric structure at the low concentrations encountered under polymerization conditions.

The above process, Eq. (25), is in conflict with the currently available theoretical results (Table 5) regarding the dissociation enthalpies of aggregated organolithiums. A similar conclusion was reached by Brown in 1966 [70]. This assessment is fortified by the fact that the measured [94] energy of activation for the reaction of styrene with n-butyllithium, 18 kcal/mole, is a value far lower than that required if the calculated dissociation enthalpy of the n-butyllithium aggregates is included in the overall energetics of the initiation event. Thus, it would seem that any mechanism which involves *only* unassociated organolithiums as reactive entities is invalid.

However, a dissociation process parallel to that shown in Eq. (2) should, at least at this juncture, not be discounted. A similar assessment can be made for the process given in Eq. (3) if alkyllithiums aggregated as dimers are reactive initiators. It is also germane to mention that the calculations of Graham, Richtsmeier and Dixon [18] show that the bonding in the hexamer of ethyllithium can consist of closed three-center Li—C—Li bonds with significant donation from a third lithium. In other words, the organolithium hexamers may be described as composed of weakly interacting trimers. Thus, the following association-dissociation equilibrium may play a role in the reaction involving styrene and n-butyllithium in benzene.

$$(n\text{-}C_4H_9Li)_6 \rightleftharpoons 2(n\text{-}C_4H_9Li)_3 \tag{26}$$

It should also be mentioned that the work of Graham and coworkers [18] apparently demonstrates that the identity of the alkyl substituent on the alpha carbon atom of alkyllithiums has little influence on the energetics involved in the various dissociation steps and the stability of the various aggregated structures.

The use of aliphatic solvents causes a profound change, for example, in the kinetic behavior of the initiation step of styrene by organolithium species, i.e., the inverse correspondence between reaction order and degree of organolithium aggregation is no longer observed. This can be seen by an examination of the findings collected in Table 10. There it can be readily seen that the use of an aliphatic solvent leads to kinetic orders which are unrelated, at least in a direct fashion, to the aggregation state of the initiating organolithium. Also, initial rates of initiation in aliphatic solvents were found to be several orders of magnitude less than those observed, under equivalent conditions when the aliphatic solvent was replaced with benzene. Pronounced induction periods are also shown by these systems.

It, thus, would appear that in aliphatic solvents the initiation process involves reaction of the monomer with aggregated organolithium species. A further complicating feature is the presence of cross-associated structures involving the initiating organolithium and the newly formed benzylic- or allylic-lithium species. Solution viscosity measurements have shown [45, 100, 103, 104] that poly(isoprenyl)lithium can form cross-associated complexes with n-butyllithium, sec-butyllithium, tert-butyllithium and ethyllithium. For the latter organolithium, evidence exists [104] that the cross-associated state is favored over the two homoaggregated species.

Kinetic complications are kown to exist when mixed aggregates are present. When 2,3-dimethyl-1,3-butadiene in heptane is initiated by n-butyllithium, the conversion-

Table 10. Initiation Studies Involving Organolithium in Hydrocarbon Solvents

Monomer	Initiator	Degree of Association[a]	Solvent	Reaction Order	Ref.
Styrene	n-C_4H_9Li	6	Benzene	0.16	94)
	n-C_4H_9Li	6	Cyclohexane	0.5–1.0	51)
	s-C_4H_9Li	4	Benzene	0.25	30)
	s-C_4H_9Li	4	Cyclohexane	~1.4	30)
2,4-Dimethylstyrene	n-C_4H_9Li	6	Benzene	0.16	95)
			Cyclohexane	0.57	95)
p-tert Butylstyrene	s- and t-C_4H_9Li	4	Benzene	0.25	96)
	s- and t-C_4H_9Li	4	Hexane	1.0	97)
1,1-Diphenyl-	n-C_4H_9Li	6	Benzene	0.16	97)
ethylene	t-C_4H_9Li	4	Benzene	0.25	98)
1,3-Butadiene	n-C_4H_9Li	6	Cyclohexane	0.5–1.0	51)
2,3-Dimethyl-1-3-butadiene	s- and t-C_4H_9Li	4	Benzene	0.25	99)
	s- and t-C_4H_9Li		Hexane	1.0	99)
Isoprene	n-C_4H_9Li	6	Cyclohexane	0.5–1.0	49)
	s-C_4H_9Li	4	Benzene	0.25	30)
	s-C_4H_9Li	4	Cyclohexane	0.75	30)
	s-C_4H_9Li	4	Hexane	0.70	100)
	s-C_4H_9Li	4	Cyclohexane	0.66	101,102)
	t-C_4H_9Li	4	Cyclohexane	0.2–0.7	101,102)

[a] RLi

time curves exhibit pronounced sigmoidal character and show an induction period [105]. Also, the propagation rate (at 30% conversion) was found to depend on total organolithium concentration raised to the power —0.32. This negative order was said to arise from the reduced reactivity of the cross-associated species $(BuLi)_x \cdot (DMBLi)_y$ when $x > y$. This interpretation was supported by the observation of a reduction in the propagation rate of 2,3-dimethyl-1,3-butadiene (as initially measured with no unreacted initiator present) on the introduction of n-butyllithium.

An analogous scenario can be observed for the reaction of styrene with purified t-butyllithium [44, 106-108]. For example, the reaction pattern in benzene revealed [108] that rapid initiation took place involving a fraction of the t-butyllithium; which was commercial material purified by sublimation under vacuum. The remaining initiator was found to react very slowly with styrene to the extent that initiator remained at the finish of the polymerization. An indication of this trend can be seen in the chromatogram of Figure 8a for initiation in cyclohexane.

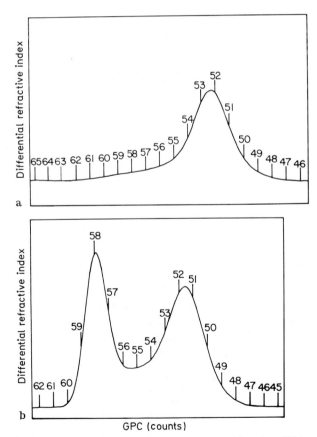

Fig. 8a. Chromatogram of polystyrene prepared by t-butyllithium in cyclohexane at 35 °C. Initial monomer concentration of 0.5M; **b.** Chromatogram of polystyrene prepared by a second monomer addition to the sample shown in Figure 8a. Monomer concentration of 0.5M after the second addition of monomer

The presence of residual initiator in the polymerization leading to the polystyrene whose distribution is shown in Fig. 8b was verified as follows [108]. At the completion of the polymerization, both THF and monomer were added to the system. The intensity of the absorption band of poly(styryl)lithium (λ_{max} 334 nm) was found to increase to an extent which demonstrated that ca. 75 % of the added t-butyllithium remained at the completion of the first polymerization.

These results indicated that the conventional spectroscopic method of determining the completion of the initiation reaction is fraught with potential uncertainties, insofar as the t-butyllithium-styrene system is concerned. Hsieh's method [106] of analyzing for residual initiator (via GC measurements for isobutane), is unencumbered with such uncertainties.

It has been suggested [107] that the failure of t-butyllithium to react completely with styrene is due to an impurity (undefined). An alternative explanation of this retarded reactivity could focus on the structure, and energetics, of the cross-associated complex (or complexes) formed between t-butyllithium and styryllithium. It is, though, currently impossible to quantitatively define the role that such cross-associated organolithium species play in these initiation events. For example, little is known regarding the stoichiometry of the cross-associated species, with the possible exception of the poly(isoprenyl)lithium-ethyllithium system where concentrated solution viscosity measurements were interpreted [104] as showing the presence of a complex of composition $PILi \cdot (C_2H_5Li)_3$. Furthermore, the bonding energetics of these cross-aggregated structures involving the allylic or benzylic active centers with alkyllithiums have not yet been examined. The formation of such cross-associated structures is not limited to hydrocarbon solutions. Burley and Young [109] have presented spectroscopic evidence that n-butyllithium and 1,3-diphenyl-1-butenyl-lithium can form such a complex in ethereal solvents, even though the latter organolithium does not undergo self-association in such solvents.

An interesting extension of the work of Worsfold and Bywater [94] is that of O'Driscoll, Ricchezza and Clark [110]. Their approach consisted of a kinetic evaluation of the reaction of n-butyllithium with styrene in benzene where the monomer concentration ranged from 0.2 to 1.0 mol l^{-1}. The maximum concentration of styrene used by Worsfold and Bywater [94] was ca. 0.03 mol l^{-1}. O'Driscoll and co-workers [110] found that the initiation rate was about an order of magnitude larger than that reported by Worsfold and Bywater [94] for comparable concentrations of n-butyl-lithium. Also, the kinetic order was found to be 1/3 in initiator at a styrene concentration of 0.5 mol l^{-1}. It was concluded that styrene stabilizes n-butyllithium in the trimeric form. Calorimetric measurements indicated that the corresponding solvation enthalpy was 10.2 $\pm$ 1.8 kcal mol^{-1} of hexameric n-butyllithium. The proposal that styrene can stabilize n-butyllithium trimers is in accord with the previously mentioned conclusion of Graham and co-workers [18] that n-alkyllithium hexamers can consist of weakly interacting D_{3h} trimers. The potential for the interaction of styrene with the n-butyllithium hexamer is also supported by the UV results of Oliver and co-workers [111,112] which showed that lithium-π electron interactions can occur between olefinic double bonds and the lithium counter-ion.

The validity of the concept that only unassociated organolithiums are reactive in initiation is predicated on an assumption that the associated and dissociated species are in *rapid* equilibrium with one another. This, of course, would result in the exchange of the organolithium molecules. The proclivities for exchange for t-butyllithium have been studied [6, 72] in cyclopentane via ^{13}C-NMR and mass spectrometric measurements. Following the removal of solvent, the mass spectra of mixtures of isotopically pure t-butyllithium-6 and t-butyllithium-7 demonstrated that complete equilibration had not occurred after 12 hours at room temperature. Thus, at least for this organolithium, a dynamic equilibrium similar to that shown in Eq. 25 does not exist. Nonetheless, as has been mentioned, t-butyllithium is, at least for dienes, an effective initiator. The implication that can be drawn from these results is that aggregated, e.g., tetramers and dimers, organolithiums can react directly with monomers. It should also be noted, that in contrast to t-BuLi, other alkyllithiums seemingly undergo fairly rapid intermolecular exchange in hydrocarbon solvents [6].

An examination of the sources for the initiation results in Table X reveals that those studies were done under conditions where both reactants were present at low concentration, $< 10^{-2}$ M. Thus, none of the results were obtained under conditions encountered in the preparation of high polymer, e.g. $[RLi]_0 = 10^{-3}$ and $[M]_0 = 1$ molar. The studies which are available [106, 107, 113–115] where initiation was measured under polymerization conditions offer a distinctly different perspective of the initiation event, relative to what has been observed for the conditions under which the findings reported in Table 10 were obtained.

Hall [113], Hsieh [106], Roovers and Bywater [107], Tanlak and co-workers [114], and Bordeianu and co-workers [115] followed the initiation of styrene under polymerization conditions in aromatic or alkane solvents using ethyllithium, i-propyllithium, or isomers of butyllithium. Without exception, these authors found a first power dependency of initiation rate on total active center concentration. Hsieh's results [106] and those of Roovers and Bywater [107] also indicate that the first order character for initiation is independent of the degree of association (4 or 6) of the alkyllithium. The first order dependence of the initiation step on total active center concentration is also maintained over the period where cross-aggregated structures, $PSLi \cdot (RLi)_x$, are present.

It could be argued that the results of Hsieh [106], Tanlak and co-workers [114] and Bordeianu and co-workers [115] were influenced by 'impurities' since their measurements were done under nonvacuum conditions. However, the work of Hall [113] and that of Roovers and Bywater [107] was done using evacuated reactors. An important contribution from the work of Roovers and Bywater [107] is their additional finding that isoprene initiated by t-butyllithium showed the near first-order dependence of initiation rate on total active center concentration in benzene at monomer concentration of $> 5 \times 10^{-2}$ molar while measurements made at isoprene concentrations of 10^{-3} to 10^{-4} molar resulted in the usual $^1/_4$ dependency.

These combined findings, taken in concert with those of O'Driscoll [110], show that even in aromatic solvents the inverse relationship between organolithium aggregation state and reaction order of the initiation step is not always observed. Excess monomer (relative to initiator) can influence the initiation event; possibly through π-complexation [111, 112] with the organolithium aggregates. Consistent with this view is that the apparent energy of activation for the initiation of styrene by n-butyllithium in aromatic solvents decreases (18 kcal/mole [94] to 6.3 kcal/mole [115]) as the monomer concentration is increased relative to the organolithium concentration. Monomer interaction with associated organolithiums was involved in a mechanism advanced by Brown in 1965 [5].

The foregoing information demonstrates that no single process, such as that shown in Eq. (24), can be invoked to describe the initiation event in hydrocarbon solvents, particularly under conditions leading to the preparation of high molecular weight polymer. Obviously, aggregated organolithiums can react directly with olefinic and diolefinic monomers. This is, perhaps, to be expected since there does not exist any obvious or compelling feature of the bonding patterns in organolithium aggregates that indicates that these species by necessity should be unreactive toward olefinic or diolefinic monomers. The concept that associated organolithiums are reactive in their own right has previously been advanced by Brown [5, 70] and others [35, 41, 111, 116−118].

6 Chain Propagation Involving Carbon-Lithium Active Centers

The allylic- and benzylic-lithium active centers, which can be characterized as having polarized covalent carbon-lithium bonds in hydrocarbon solvents, have been extensively studied with regard to their structure, their kinetic behavior in the propagation event, and their association states. These latter two topics are the subject of this Section.

It is recognized that the propagation reaction involving dienes or styrene is dependent upon active center concentration raised to a fractional power; 1/2 in the case of styrene [38, 51, 94, 119] and 1/4 to 1/6 for the 1,3-butadiene [51, 120−122], 2-methyl-1,3-butadiene [49, 121−124] and 2,3-dimethyl-1,3-butadiene [105]. Unlike the situation encountered in the initiation event of styrene and dienes, the reaction order dependence of the propagation process on active center concentration is *independent* of the identity of the hydrocarbon solvent, aromatic or aliphatic, although the relative propagation rates, under equivalent conditions, are faster in benzene than an aliphatic solvent.

Although data are available [125−130] for dienes which indicate a 1/2 order dependence of the propagation rate on active center concentration, it appears that the lower orders are the correct exponents for these reactions. As a consequence, it has been proposed [49] that a process similar to that shown in Eq. (25) holds for these systems, i.e.,

$$(\text{PILi})_4 \rightleftharpoons 4 \text{ PILi} \tag{27}$$

where only the unassociated poly(isoprenyl)lithium species participate in the propagation step. Shamnin, Melevskaya, and Zgonnik [131], believe that the aggregated forms of poly(butadienyl)lithium participate in propagation and in particular that the dimeric associate is responsible for 1,2-enchainment.

The association states of the benzylic- and allylic-lithium active centers have been studied by viscosity, light scattering and cryoscopy (Table 2). The majority of results indicate that the dimeric state of association is present for these active centers at the concentrations appropriate for polymerization (10^{-3} to 10^{-4} M).

The viscometric method relies on the relation (in the entanglement regime):

$$\eta = KM_w^{3.4} \tag{28}$$

where the constant K includes the concentration term. Since the concentration of the polymer solution remains virtually unchanged after termination of the active centers, the foregoing relation can be modified as follows:

$$\frac{\eta_a}{\eta_t} = \frac{t_a}{t_t} = \left[\frac{M_{w_a}}{M_{w_t}}\right]^{3.4} = N_w^{3.4} \tag{29}$$

where t denotes the polymer solution flow time, the subscripts a and t the active and terminated solutions, and N_w the weight average association number. The viscometric method potentially represents a convenient and accurate method whereby active center aggregation can be measured.

The technique involving an evacuated capillary viscometer is limited to the measurement of polymer solutions having viscosities less than ca. 10^3 poise; a point which has been made previously by Hadjichristidis and Roovers [132]. This is a consequence of the viscometer type (Ubbelohde) which has been used in these determinations. Despite this practical limitation, it has been stated [78] that "there is no inherent limit for viscosity measurements, even in conventional types of viscometers provided that the tubes are sufficiently wide." However, the limitations of operating with an evacuated viscometer, and the flow behavior of high viscosity ($> 10^3$ poise) polymer solutions, clearly reveals that the foregoing claim is unrelated to reality.

An examination of the results in Table II shows that the light scattering results of Worsfold and Bywater [43,49] yield association numbers that are larger than those reported by other groups. Worsfold [133] has presented results which are claimed to demonstrate that the viscosity technique will not detect the presence of poly-(dienyl)lithium aggregates with association states greater than two. Briefly, his model can be described by considering linear chains with associating groups at one end. Such chains will spend some time in the monomeric form and some time in the associated state, e.g., as a dimer. As the energy of association increases, the rates of interconversion (monomer $\rightleftharpoons$ dimer) will decrease. If these rates are slow enough, equilibrium measurements (osmometry and light scattering) should give the same information as dynamic measurements (viscosity and diffusion coefficients). Differences are possible if the interconversion rates are rapid, i.e., if the dimer lifetime is relatively short.

Thus, Worsfold [133] measured the association behavior, via light scattering, UV-visible spectroscopy and viscosity, of polystyrene chains capped at one end with the dimethylamine group where the associating group was the bis-(2,6-dinitrohydro-quinol). This bidentate species;

can complex with the tertiary amine groups via the phenol groups. Although not considered by Worsfold, the above compound can also self-associate by interaction of the phenol groups either with themselves or with the carbonyl oxygens.

Worsfold found that the degree of association as measured from viscosity was less than that indicated by the light scattering and spectroscopic results. It was therefore concluded that the association ⇌ dissociation rates were comparable to the chain entanglement lifetime. As a consequence, Worsfold concluded that viscosity measurements involving concentrated solutions of poly(dienyl)lithium in the entanglement regime could not detect the presence of, for example, star-shaped tetramers if the equilibrium

$$(PILi)_4 \rightleftharpoons 2 \ (PILi)_2 \tag{30}$$

is very labile.

Worsfold [133] reaffirmed that the true state of association is four for poly-(butadienyl)lithium while poly(isoprenyl)lithium is a mixture of dimers and tetramers [43]. These conclusions, though, were presented without reference to the light-scattering and cryoscopic studies (Table 2) which indicate that the dienyllithium active centers can assume the dimeric state at concentrations encountered in the preparation of high molecular weight polymers. A pertinent example is the cryoscopic results of Glaze and co-workers [36] which showed that neopentylallyllithium (the reaction product of 1,3-butadiene and t-butyllithium) assumes the dimeric association state at low (<0.1 m) concentrations. These results of Glaze are in general accord with those of Makowski and Lynn [41] who also found (via viscosity measurements) that the poly(butadienyl)lithium active center can exhibit a variable degree of aggregation.

Furthermore, several of Worsfold's assessments seem to be open to question. The assertion that "the association (between the allylic-lithium active centers) is between ionic species" can be contrasted with the evidence provided by NMR spectroscopy [36, 134−143] which has shown that the carbon-lithium bond of allylic-lithium species can possess considerable covalent character. Worsfold has also previously published [43] concentrated solution viscosity results where the ratio of flow times, before and after termination, of a poly(isoprenyl)lithium solution was about 15. This finding is clearly incompatible with the conclusion that viscometry cannot detect the presence of aggregates greater than dimeric.

It should also be noted that the viscometric technique can detect the presence of star-shaped aggregates, having the ionic active centers. The addition of ethylene oxide to hydrocarbon solutions of poly(isoprenyl)lithium leads to a nearly two-fold increase in viscosity [144]. Conversely, this results in an approximately twenty-fold decrease in solution viscosity after termination by the addition of trimethylchlorosilane. This change in solution viscosity is reflected in the gelation which occurs when difunctional chains are converted to the ionic alkoxy active centers [140, 145, 146]. Branched structures have also been detected [147] by viscometry for the thiolate-lithium active center of poly(propylene sulfide) in tetrahydrofuran.

The alteration in solution viscosities brought about by the conversion of the allyllic-lithium active center to the alkoxy-lithium species is in accord with the general trend [148, 149] observed for star-shaped polymers in concentrated solution. It must be noted though that viscosity measurements cannot generally be used to detect differences in the degree of branching of star-shaped polymers for the case where the arm molecular weight remains constant while the extent of branching is changed. This facet of star-shaped polymer rheology has been demonstrated by experiment [148,149] and theory [150].

Recent semi-quantitative solution viscosity measurements by Hsieh and Kitchen [151] using a viscosity monitoring device in a reactor in combination with end-capping and linking reactions suggest that the poly(dienyl)lithium active centers (where poly(butadienyl)lithium > poly(isoprenyl)lithium) are more highly associated than poly(styryl)lithium. These results could be taken as a demonstration that the effective degree of association of polymeric organolithiums may be dependent on the time scale of the technique used for the measurement. Worsfold and Bywater [43] have proposed a mechanism to explain such apparent anomalies in terms of intermolecular, inter-aggregate exchange reactions.

However, Hsieh and Kitchen [151] failed to consider the influence of their measurement temperature, 78 °C, on the stability of the poly(dienyl)lithium active centers (see section on Active Center Stability). As an example of this potential problem is the observation by two separate groups [47,152] that viscometric measurements of hydrocarbon solutions of poly(butadienyl)lithium fail to yield constant flow times (at 30 °C) following the completion of the polymerization, i.e., the flow times were found to increase with increasing time. This inability of the poly(butadienyl)lithium chain to exhibit constant solution viscosities renders it unsuitable for association studies of the type done by Hsieh and Kitchen [151].

It is becoming more widely recognized that the kinetic consequences of the aggregation involving carbon-lithium species are but imperfectly understood regarding both the initiation and propagation processes. An example of this can be seen for 1,3-butadiene. Johnson and Worsfold [51] have shown that the propagation rate exhibits a kinetic order of 1/6 with regard to active center concentration. However, no direct relation between kinetic order and association state exists; even if the tetrameric association state advanced by Worsfold and Bywater [43] is accepted.

The influence of the interaction of organolithium active centers with π-electron donors on the propagation kinetics remains unelucidated. These species can include such weak donors as durene [153] and lithiated 1-hexyne [154]. The formation of such a complex between styrene and poly(styryl)lithium dimers has been invoked by Kaspar and Trekoval [155]. In essence, they proposed a mechanism involving the slowly established equilibrium formation of monomer-complexed poly(styryl)lithium dimers [155] in order to explain their kinetic results for the styrene system where the order of the propagation reaction in cyclohexane was found to approach unity at an active center concentration of ca. 10^{-2} M. They suggested that the complex takes the form of a tetragonal bipyramid having at its apices lithium atoms with a coordination number of four. The base of the pyramid was envisioned as consisting of two sp^2 hybridized orbitals of the styrene vinyl group α and β carbons and the two sp^2 hybridized carbanions of the poly(styryl)lithium dimer. It is also germane to mention that π-electron interactions have been suggested [143] to occur between the double bonds in poly(1-butenylene) and the aggregated active centers; a finding in accord with that of Smart and co-workers [110,111] for 3-butenyllithium (which was found to possess the hexameric association state in cyclopentane [111]). It is appropriate to again mention that a mechanism involving monomer complexing with aggregated organolithium active centers as the rate determining step was advanced by Brown[5] in 1965.

7 Polymerization in the Presence of Ethers

Remarkably few systematic studies have been made of the kinetics of anionic polymerization in non-polar solvents containing small amounts of ethers; in contrast, studies of bulk ether systems abound. Several studies have appeared [156-158] in which the propagation reactions involving styryllithium were measured in mixtures of benzene or toluene with ethers. The kinetic orders, in some cases, of the reactions were identical to those observed in the absence of the ether. Thus, in part, the conclusion was reached [157,158] that the ethers did not disrupt the dimeric degree of aggregation of poly(styryl)lithium. The ethers used were tetrahydrofuran [156], anisole [157], diphenyl ether [158], and the ortho and para isomers of ethylanisole [157].

An early investigation is that of the influence of tetrahydrofuran (THF) upon the polymerization of styrene in benzene by Bywater and Worsfold [156]. For small ratios of THF: active centers there is a marked increase in rate; but at ratios above about 10:1 the rates decline. The absorption spectrum is unchanged by the presence of small proportions of THF (2:1) but some broadening is observed at higher concentrations. It was proposed that two complexes are involved — a rather reactive monoetherate and a much less reactive dietherate; although, it is not obvious why these species should have these relative reactivities. When large amounts of THF are present (0.15 molar) the propagation exhibits a first order dependence upon chain end concentration but with smaller amounts (up to THF/active centers $\cong$ 17) a square root dependence is obtained. These results were interpreted to imply that with little THF most of the chain ends form unreactive dimeric associates in equilibrium with two reactive species — the non-associated chain ends in the monoetherate and non-solvated forms; at high levels of THF all association is removed. Investigation of the influence of THF upon the concentrated solution viscosity of poly(styryl)lithium dimers revealed [42] that disruption of association occurs at low levels of THF (the weight-average degree of association falling from 2.0 to ca. 1.5 on introducing a level of THF: active centers of 2:1).

In an analogous study, Geerts, Van Beylen and Smets [157] found that the propagation of styrene also exhibited a square root dependence of rate upon active center concentration when anisole or 4-ethylanisole was present over an ether/lithium range of 33 to 1030, and also in the presence of 2-ethylanisole for the range 40 to 3980. The cross-over reaction between poly(styryl)lithium and "double diphenylethylenes" in benzene with diphenyl ether (ether lithium of 150) also showed a one-half order dependence on active center concentration [158]. The influence of tetrahydrofuran, diphenyl ether and anisole on the association of poly(styryl)lithium active centers in benzene solution has been examined [42,52] by the viscometric technique. The results revealed that these ethers can cause disaggregation. The relative influence of these ethers on poly(styryl)lithium aggregation was found [52] to be as follows: tetrahydrofuran ($K \cong 160$) > > > anisole ($K \cong 4.2 \times 10^{-2}$) > diphenyl ether ($K \cong 1.4 \times 10^{-2}$) (units in LM^{-1}) where K is the equilibrium constant for the process outlined as Eq. (11). Thus a comparison of the association numbers found from the viscosity measurements [42,52] with the invariancy of the kinetic orders from the value one-half reveals that no simple connection between these parameters exists for these ether modified systems.

It should be mentioned that the viscometric technique yields a weight-average degree of association [159], N_w, while a number-average (mole fraction) degree of association is actually required for the calculation of equilibrium constants associated with processes such as shown in Eq. (11). However, as is shown in the Appendix the near-monodisperse nature of these polymer systems permits the use of N_w in the calculation of these equilibrium constants.

The validity of the viscosity measurements regarding the reported [52] influence of anisole and diphenyl ether on the association of the poly(styryl)lithium dimers has, though, been questioned [78,160,161]. Suffice it to note that the fallacies in the data provided [160,161] have been commented upon elsewhere [162]. Even though it is well-known that ethereal solvents can interact with organolithium compounds, no explanation was given [78,160,161] as to why aromatic ethers should be completely exempt from this general behavior.

The opinion has also been expressed [133] that the viscometric technique will "overestimate the degree of dissociation" in systems where the dissociation constants involving the influence of ethers are studied. This claim can be examined by considering a simple example. Bywater and Worsfold [156] and Meier [79] studied the influence of tetrahydrofuran on the propagation rate of styrene in benzene. Their kinetic results can be interpreted as showing that at an ether/active center ratio of about 10, the poly(styryl)lithium dimers were largely disrupted by solvation with the ether for the process shown in Eq. (11). This joint conclusion [79,156] is identical to that reached by Morton [42] via the viscometric technique. Thus, at least for the case of poly(styryl)lithium, the viscometric procedure does not appear to overestimate the extent of dimer dissociation.

Szwarc and Wang [161] have claimed, without citing any supportive experimental measurements, that the equilibrium constant for the process shown in Eq. (11) is approximately one; a value in stark contrast to that of ca. $160 \, LM^{-1}$ reported by Morton. [42] Their assessment is not supported by the viscometric measurements, [42] the kinetic findings of Bywater and Worsfold, [156] the association measurements reported by West and Waack [56] (Table 3) for benzyllithium in tetrahydrofuran, nor by the findings of Kminek, Kaspar, and Trekoval [161a].

An example follows. For the case where the concentration of THF was ca. 4×10^{-2} M and the active center concentration 1.2×10^{-3} M, the interpretation given by Bywater and Worsfold [156] to their kinetic findings shows that the propagation event is carried solely by ether complexed active centers, i.e., the population of associated poly(styryl)lithium dimers and unsolvated poly(styryl)lithium active centers is negligible. In contrast, the Szwarc-Wang [161] value of one for the equilibrium constant of Eq. (11) dictates that at least 45 percent of the active centers exist in the self-associated state. Morton's viscometric findings are also supported by the conclusion reached elsewhere [163] that "a 20-fold excess [of THF relative to active centers] suffices to dissociate all the lithium polystyryl."

A potentially valuable contribution to our understanding of the generality of a connection between the order of the propagation and active center concentration could have come from the study [164] of the polymerization of o-methoxystyrene in toluene where the results were the described as demonstrating that "at the higher concentration the ion-pairs occur predominately in the associated form while increasing amounts of free-ions are present at lower concentrations." However, as has been previously mentioned [162], the claimed non-linearity of their plot [76a, 164] is incorrect since a least mean squares analysis shows that the gradient is 0.62 with a 0.9994 correlation coefficient (Fig. 9). Thus if the claimed connection between the degree of aggregation and the observed kinetic order is to hold, the association state of the styryl active centers would be constant over the same concentration range $(1.8 \times 10^{-2}$ to 5.3×10^{-4} M). Acting on the assumption that the aggregated chain ends were incapable of growth, Smets and co-workers [164] used a curve fitting procedure to deduce that the dissociation equilibrium constant was 10^{-3} molar. This value dictates that the percentage of unassociated active centers would vary from about 15 to 61 over the concentration range studied. Clearly, these values are incompatible with those outlined above.

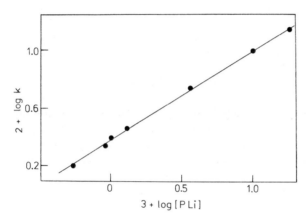

Fig. 9. Rate of polymerization vs. active center concentration for 2-methoxystyrene. Rate units are in minutes. (Reprinted with permission from Ref. [52], Copyright 1981, American Chemical Society)

It is of interest to note that Kminek, Kaspar and Trekoval [165] have made calorimetric measurements on the interaction of *n*-butyllithium with several Lewis bases at RLi/base ratios of one. The largest enthalpy changes (in agreement which the results of Quirk and Kester [80-82]) were found for TMEDA and dimethoxyethane and the smallest with diethyl ether and anisole. Their results for anisole clearly show that even aromatic ethers will interact and complex with organolithium species. Thus, their findings serve to fortify the viscometric findings regarding the influence of aromatic ethers on the poly(styryl)lithium association state.

It appears that for the ether-modified systems based on styrene, a direct relationship between the extent of active center aggregation and kinetic order is not necessarily a common feature, attractive as such a concept may be. Thus a single mechanism involving only unassociated organolithiums as the sole reactive species should perhaps not be invoked as the only explanation for the propagation reactions involving styryl and diene monomers. Both Smart [112] and Bartlett [116] have presented findings which involve aggregated organolithiums as reactive entities in the presence of ethers and tertiary amines.

The influence of tetrahydrofuran on the propagation and association behavior of poly(isoprenyl)lithium in n-hexane has been examined [47]. As for the case of poly(styryl)lithium [156], the rate of polymerization was found to first increase followed then by a decrease as the THF/active center ratio increased. This decrease ultimately reached the polymerization rate found in pure tetrahydrofuran at a THF: active center ratio of ca. 2×10^3. This was for the case where the active center concentration was held constant and the tetrahydrofuran concentration varied. The maximum rate of polymerization was found to occur at a THF:active center ratio of about 500; a value at which the viscometric measurements demonstrated [47] the virtual absence of poly(isoprenyl)lithium self-aggregation. As noted before in this review, the equilibrium constant for the process shown in Eq. (12) has the relatively small value of about 0.5 LM^{-1}, which is in sharp contrast with the value of about 160 LM^{-1} found for the THF-poly(styryl)lithium system. The possibility of complexation of THF directly with the poly(isoprenyl)lithium aggregates, Eq. (13), was not considered by Morton and Fetters [47].

Davidjan et. al. [166] have made a study of the influence upon the propagation of poly(isoprenyl)lithium in n-hexane of very small additions of 1,2-dimethoxyethane (DME:active centers = 0.01). Analysis of polymer obtained at 10% conversion by size exclusion chromatography coupled with a determination of the dependence of the stereochemistry upon molecular weight led them to the conclusion that complexation reduces the reactivity of what they assumed to be the most reactive species, i.e., the non-associated active center.

A clear consensus [47, 156, 166] has emerged which indicates that various extents of ether complexation with active centers can reduce their reactivity in the chain propagation event. If cation-monomer coordination is important, the presence of ether in the coordination sphere might be expected to lead to less monomer interaction with a subsequent reduction in polymerization reactivity. Clearly, there is a need for further work, experimental *and* theoretical, on this topic.

8 Polymerization in the Presence of Tertiary Aminees

Since the linear lithium alkyls can be regarded as living oligomers of ethylene, it might be supposed that these would propagate under suitable conditions. An early study by Ziegler and Gellert showed [167] that when propyllithium reacts in ether with ethylene overnight and is subsequently treated with formaldehyde, a mixture of C6, C8, C10 and C12 alcohols is obtained. Langer [168] found that alkyllithiums can polymerize ethylene to high molecular weight linear high density poly(methylene) in the presence of N,N,N',N'-tetramethylethylenediamine (TMEDA) and proposed that the active species has the chelated structure.

Eberhardt et al. [169,170] found that TMEDA, sparteine or other ditertiary amines enable the telomerization of ethylene with benzene by lithium alkyls to yield molecules of the general formula $C_6H_5(CH_2CH_2)_nH$. A valuable compilation has been made of related studies employing polyamine chelated alkali metal compounds [87].

The first kinetic studies were made by Hay and coworkers [171,172]. They found that the rate of polymerization of ethylene was independent of the concentration of TMEDA and concluded that the active initiating species is n-butyllithium which is neither complexed nor self-associated; initiator efficiences were reported to be less than 50%. The rate of consumption of ethylene was found to be proportional to the concentrations of ethylene and n-butyllithium.

Magnin and coworkers [173] also studied the kinetics of the polymerization of ethylene in hexane using n-BuLi/TMEDA as initiator and obtained results very different from those of Hay et al. [171,172]. The reaction was found to be first order in ethylene and to exhibit a square root dependence upon whichever of the substances TMEDA or RLi was present in the smaller quantity. With a constant concentration of organolithium, the rate increased on increasing the concentration of TMEDA until a limiting value was reached when [TMEDA] = [RLi]. These observations were rationalized by the scheme:

$$(RLi)_n + nTMEDA \rightleftharpoons \frac{n}{2}[RLi \cdot TMEDA]_2 \tag{31}$$

$$(RLi \cdot TMEDA)_2 \rightleftharpoons 2(RLi \cdot TMEDA) \tag{32}$$

$$(RLi \cdot TMEDA) + C_2H_4 \rightarrow (RCH_2CH_2Li \cdot TMEDA) \tag{33}$$

The addition of TMEDA to the system results in the formation of a complex of 1:1 stoichiometry which is present largely as an unreactive dimer, in equilibrium with a small amount of the highly reactive monomeric complex. Under these conditions the overall rate of polymerization R_p is given by

$$R_p = k_p [\text{Ethylene}] K^{1/2} [(RLi \cdot TMEDA)_2]^{1/2}$$
$$\cong k_p K^{1/2} [\text{Ethylene}] [X]_0^{1/2} \tag{34}$$

where the amount of complex is limited to whichever of the initial concentrations of TMEDA or RLi is the smaller. In an earlier piblication, the same group [174] reported that the consumption of the initiator was complete even if the ratio of [TMEDA]:[RLi] was less than one. This implies that the exchange equilibrium

$$[RLi \cdot TMEDA] + R'Li \rightleftharpoons [R'Li \cdot TMEDA] + RLi \tag{35}$$

is fast, compared to the propagation process. At the very high concentration of alkyllithium employed by Hay et al. (0.5 to 1 molar), Magnin et al. [173] found that their polymerization rate showed a steeper dependence of rate upon [RLi] than the onehalf power they observed at lower (0.03 to 0.3 M) concentrations. As they remarked, with [BuLi] = [TMEDA] = 1 M, some 30% (v/v) of the reaction mixture is TMEDA-a situation very different from that at lower concentration.

One feature of the earlier work of Schue et al. [174] that does not fit their subsequently proposed mechanism [173] is their observation that the propagation rate is directly proportional to the initiator concentration when [TMEDA] = [t-BuLi]. They did not discuss this point in their latter publication [173].

All three isomers of butyllithium in the presence of TMEDA give very similar results for the propagation reaction. With n-butyllithium the initiation process proceeds at the same rate as propagation, but with s- and t-butyllithium the initiation is faster than propagation [173–175]. In these last two cases, the process of initiation converts the very reactive secondary and tertiary carbanions into the primary ion. A similar phenomenon has been reported by Bartlett et al. [176] who found that i-propyllithium in ether solution at −60° adds only a single molecule of ethylene.

Crassous et al. [177] studied the polymerization of ethylene using n-butyllithium in conjunction with the tertiary diamines TMEDA, TEEDA (tetraethylethylenedi-amine) and PMDT (pentamethyldiethylenetriamine). In contrast to the situation with TMEDA the rate of polymerization was found to show a first order dependence upon the complexed chain end concentration. Steric hindrance seems to prevent the dimerization of the chain ends. Examination of the n-BuLi. TEEDA complex by ^{1}H-NMR shows that the displacement to high field of the protons α to the lithium induced by complexation is much smaller with TEEDA than with TMEDA or PMDT. With [TEEDA] < [RLi] time-averaged signals were obtained showing that the exchange process

$$(RLi \cdot TEEDA) + R'Li \rightleftharpoons (R'Li \cdot TEEDA) + RLi \qquad (36)$$

is fast on the NMR time-scale. A direct comparison of the propagation rate constants for poly(methylene)lithium is not possible in the absence of a value for the dissociation constant for the process:

$$(RLi \cdot TMEDA)_2 \rightleftharpoons 2(RLi \cdot TMEDA) \qquad (37)$$

No detailed kinetic study was performed with PMDT because rapid metallation of PMDT by the complexed organolithium resulted in the generation of a species incapable of initiation. Metallation of TMEDA has also been described by Langer [168], but in that system initiation did subsequently occur giving rise to polymethylenes containing nitrogen. It is perhaps worth mentioning at this point that treatment of TMEDA with molecular sieve, calcium hydride, sodium or sodium-potassium alloy is effective in drying. However, treatment with n-butyllithium results in metallation to yield a product susceptible to decomposition into lithium dimethylamide and N,N-dimethylvinylamine [169] according to the scheme:

$$(CH_3)_2N(CH_2)_2N(CH_3)_2 + n\text{-}C_4H_9Li \rightarrow$$
$$(CH_3)_2NLi + CH_2{=}CHN(CH_3)_2 + C_4H_{10}$$

Clearly, purification of TMEDA by treatment with n-butyllithium, as described by Hay et al. [171], is not to be recommended.

Helary and Fontanille [92] studied the propagation of poly(styryl)lithium in cyclohexane in the presence of small quantities of TMEDA. They found (Fig. 10) that in

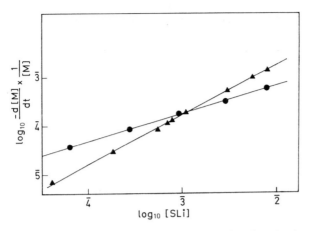

Fig. 10. Determination of the reaction order as a function of active centers at 25 °C. ● in the absence of TMEDA. ▲ for [TMEDA]/[PSLi] = 1. Rate units in sec. (Reprinted with permission from Ref. [92]), Copyright 1978, Pergamon Press)

the absence of TMEDA the propagation reaction exhibits a one-half order dependence upon chain concentration, but that when TMEDA is present at a ratio of [TMEDA]:[Li] of unity, the order becomes first. In consequence, the addition of equimolar TMEDA to a polymerizing system causes an increase in rate when the concentrations are greater than that corresponding to the intersection (ca. 10^{-3} molar at 25°); at lower concentrations the addition causes a decrease in rate. Since the temperature dependence of the rate of propagation has been determined in the presence and absence of TMEDA, it is possible to calculate the temperature dependence of the intersection. Raising the temperature to 40 °C alters the concentration at intersection to ca. 10^{-2} molar. They found that the absorption maximum of poly(styryl)lithium in cyclohexane moved from 326 nm to longer wavelengths on adding TMEDA, reaching a limiting value of 342 nm at a ratio [TMEDA]:[Li] $\geqq$ 1. They concluded that the kinetic and spectroscopic observations are in accord with the equilibria:

$$(PSLi)_2 \rightleftharpoons 2\,PSLi \tag{38}$$

$$(PSLi)_2 + 2\,TMEDA \rightleftharpoons 2\,PSLi \cdot TMEDA \tag{39}$$

and that equilibrium (38) lies far to the left and (39) far to the right. The reactivity of the complexed chain end is less than that of the uncomplexed non-aggregated chain end; a quantitative comparison cannot be made in the absence of a knowledge of the values of the equilibrium constants for processes (38) and (39). The dimeric chain ends were assumed to be unreactive.

Helary and Fontanille [178]) have studied the influence of the tertiary amine crown molecule tetramethyltetraazocyclotetradecane (TMTA) upon the propagation of poly(styryl)lithium in cyclohexane at 20 °C using a rapid calorimetric procedure. The complexing agent, employed at a [TMTA]:[Li] ratio of unity, caused a shift of the absorption maximum from 326 to 345 nm, indicating an increase in interionic separation; it was presumed that, under these conditions, there is no aggregation of the chain ends. The rate constant calculated on this basis was $750 \, l \, mol^{-1}s^{-1}$ — a value much smaller than that observed in benzene: THF mixtures by Worsfold and Bywater [179] $(40,000 \, l \, mol^{-1}s^{-1})$. The suggested interpretation [178] that the C—Li bond is stretched, increasing ionicity but without causing true separation of charges, appears to take no cognizance of the geometrically crowded lithium atom.

The influence of tertiary bases, such as TMEDA, upon the polymerization of conjugated dienes is at once more complex than that of olefins because of the variation in chain stereochemistry that accompanies the changes in rate. In an effort to simplify the discussion, the question of polymer stereochemistry is deferred to a separate Section.

The kinetics of the polymerization of butadiene by n-butyllithium in the presence of TMEDA was studied by Hay and McCabe [180]. They were unable to distinguish between addition of monomeric n-butyllithium and that of the species (n-BuLi: TMEDA) to the monomer as the initiation step. The initiation efficiency varied from 50% at a ratio of [TMEDA]:[Li] of 0.9 to 99% at a ratio of 3.35 and it was concluded that propagation involves growth from the loose (solvent separated) ion pair of composition (PBLi $\cdot$ 2 TMEDA). The presumption that there is no exchange of TMEDA among the complexed species is not in accordance with the observation of time-averaged signals in the ^{1}H-NMR spectrum [181].

Vinogradova et al. [182] found that the rate of polymerization of butadiene in petroleum ether at 20 °C reaches a maximum value when the ratio of [TMEDA]:[Li] is about 4. Measurement of the flow times of a dilute solution of high molecular weight poly(butadienyl)lithium containing an equal amount of TMEDA before and after termination, suggested that the chains are largely in the non-aggregated form. Analysis of the IR spectrum showed that the stoichiometry of the complexed chain end is RLi $\cdot$ TMEDA.

The measurement of concentrated solution viscosities affords a much more sensitive indicator of the extent of association than is provided by the dilute solution work of Vinogradova et al. [182] Using an Ubbelohde viscometer under rigorous conditions of high vacuum, Milner, Young and Luxton [152] measured the flow times of solutions of poly(butadienyl)lithium of appropriate molecular weights and concentrations before and after the addition of small amounts of TMEDA. From the flow times of the terminated solutions, the mean weight-average degree of association N_w, was determined for a range of values of r = [TMEDA]:[Li]. The results (Fig. 11) show that on the addition of TMEDA, N_w falls from its initial value of 2.0 towards 1.0 as r approaches unity. Were every original dimeric aggregate dissociated completely by the addition of two molecules of TMEDA, N_w would fall linearly with r as shown by the line in Figure 11; in fact N_w falls somewhat faster than this. A possible complication which may arise is from the formation of the 3-vinylcyclo(pentyl-l)-lithium unit for chains having a vinyl penultimate unit [183–185]. This could involve as much as about 10% of the active centers, but the cyclic structure so formed, if associated, would exhibit a dimeric or greater degree of aggregation. As has been mentioned, Worsfold [133] has suggested that the rate of association-dissociation may be altered so that chain "entanglements" may be removed by one chain slipping through the gap transiently open between associating active ends. The effect, if such a mechanism exists, would be to cause a drop in viscosity. In the present case, the reduction in the mean lifetime of a dimeric aggregate could arise from the

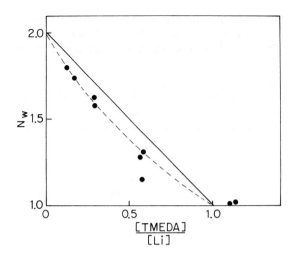

Fig. 11. Influence of TMEDA on the association of the poly(butadienyl)-lithium active center. (Reprinted with permission from Ref. [152], Copyright 1983, IPC Science and Technology Press)

exchange reaction;

$$(RLi)_2 + R'Li \cdot TMEDA \rightleftharpoons (RLi \cdot R'Li) + RLi \cdot TMEDA \qquad (40)$$

on the premise that two molecules of TMEDA complex and dissociate one dimer. It would be expected that the rate of polymerization would reach an upper (or lower) limit when $r = 1$ unless a still more (or less) reactive complex containing a greater proportion of TMEDA can be formed at still higher values of r.

The presence of TMEDA has been reported to cause a decrease in the rate of polymerization of 2,3-dimethylbutadiene [186] but an increase in that of isoprene [187].

Table 11. Analysis of Polyisoprene Fractions Recovered at Varying Conversions of Monomer to Polymer [189]

Conversion %	Fraction Number	Weight[a] Fraction (%)	Stereochemistry (%)			M expt.
			1,4	1,2	3,4	
10 (A)	I	11	69	3	28	8.5×10^4
	II	78	60	5	35	1.1×10^4
	III	11	16	7	77	5.0×10^3
35 (B)	I	19	69	3	28	
	II	73	54	6	40	
	III	8	33	10	57	
80 (C)	I	8	65	7	28	
	II	84	57	3	40	
	III	8	43	15	42	

[a] An evaluation of the chromatograms [189] upon which these values are based fails to support these weight fraction values.

Contrarily, the rate of isoprene polymerization has been found to decrease when TMEDA was added [92)188)]. If isoprene is analogous to styrene [92)], the direction of change in rate on adding TMEDA will depend upon the concentration of chain end.

In a series of experiments in which r = 0.01 Davidjan et. al. [189)] have made a systematic study of the influence of TMEDA upon the molecular weight and stereochemical distribution of poly(isoprenyl)lithium formed in hexane at −30 °C. Reaction mixtures were allowed to polymerize to 10,35 and 80% conversion before being

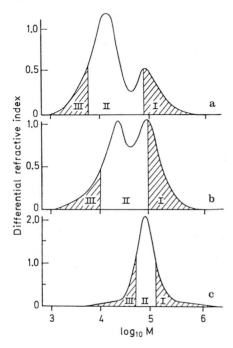

Fig. 12a–c. Chromatographs of polyisoprenes prepared in the presence of low concentrations of TMEDA. (Reprinted with permission from Ref. [189)], Copyright 1979, Hüthig and Wepf Verlag)

terminated. The products were each divided into three fractions which were analyzed by size exlusion chromatography (SEC). The results are tabulated in Table 11 and the SEC traces shown in Fig. 12. At low conversion the product has a strongly bimodal molecular weight distribution; which sharpened with increasing conversion. The stereochemistry of the polymer of fraction III obtained at low conversion is very similar to that obtained when larger amounts of TMEDA are employed and it was concluded that this fraction originated from a TMEDA complexed aggregate. The high molecular weight fraction I is distinguished by a high 1,4-content, ca. 70%, as compared to ca. 80% obtained in the absence of TMEDA. This implies that fraction I arises from growth of TMEDA-free active centers which occasionally participate in reversible complexation. The potential equilibria which must be considered follow:

$$(M_xLi)_n \rightleftharpoons nM_xLi \tag{41}$$

$$(M_xLi)_n + TMEDA \rightleftharpoons (M_xLi)_n \cdot TMEDA \tag{42}$$

$$(M_xLi)_n \cdot TMEDA + (M_yLi)_n \rightleftharpoons (M_xLi)_n + (M_yLi)_n \cdot TMEDA \tag{43}$$

$$M_xLi + TMEDA \rightleftharpoons M_xLi \cdot TMEDA \tag{44}$$

$$M_xLi \cdot TMEDA + M_yLi \rightleftharpoons M_xLi + M_yLi \cdot TMEDA \tag{45}$$

$$(M_xLi)_nTMEDA + M_yLi \rightleftharpoons (M_xLi)_n + M_yLi \cdot TMEDA \tag{46}$$

In the original paper n was taken to be four [189] although the burden of evidence suggests (Table 2) two is more likely. This scheme assumes that complexation need not result in disaggregation. Viscometric studies show that poly(butadienyl)lithium aggregates are broken on complexation [152]. Calorimetric measurements [89] on the interaction of TMEDA with poly(isoprenyl)lithium have yielded data that are not inconsistent with the formation of a complex between one molecule of the former and two of the latter.

At the low level of TMEDA used, equilibrium (45) is not likely to be of any significance but all the other equilibria must be considered. It is clear from the results of the SEC study that several species propagate and that at $-30\ °C$ the exchange among these is quite slow. Analysis of the SEC results is complicated by the curious decision of Davidjan et al. to use different monomer and organolithium concentrations in the three experiments carried to different conversions. They concluded however that the reactivity increases in the sequence

$$(RLi)_n < (RLi)_n\ TMEDA < RLi \cdot TMEDA < RLi\ .$$

Accordingly, at low conversion, fraction III is formed principally from $(RLi)_n \cdot$ TMEDA, fraction I from RLi and fraction II from all of the active species.

Cheminat et al. [190] have investigated the polymerization of isoprene using the tertiary diamines p-tetramethylphenylethylenediamine (p-TMPA), o-tetramethylphenylenediamine (o-TMPA) and bis(4-dimethylaminophenyl)methane (DMAPM). A modest increase in the initiation rate was observed for p-TMPA and for DMAPM, reaching a limiting value at r = 1.5; with o-TMPA the increase was much greater. A parallel kinetic study of the rate of addition of n-butyllithium to 1,1-diphenylethylenene in the presence of these bases was made. By assuming that the kinetic order in the butyllithium was the inverse of its degree of aggregation they concluded that the degrees of association of the 1:1 complexes were 6 for DMAPM, 4 for p-TMPA and 1 for o-TMPA. The dangers of this kind of argument are obvious. The same comment may be made concerning the conclusion that the observed first order dependence of the rate of initiation implies that initiation involves both associated and unassociated complexes of butyllithium.

9 Spectroscopic Studies of Carbanions

The electronic spectra of carbanionic polymers have been exploited for many years as a means of determining the concentration of the absorbing species. More recently, careful study has shown that other information may be obtained — notably concerning the conformation of allylic anions. Even more powerful in this respect has been the application of NMR techniques, greatly helped by the development of instrumentation capable of examining nuclei (such as ^{13}C) at the natural abundance level.

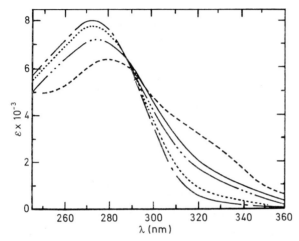

Figure 13. Absorption spectra of poly(isoprenyl)lithium in n-hexane as a function of concentration. The broken curves show increasing absorption at 320 nm in the order —·—·—, 1.9×10^{-3}; -------, 2×10^{-4}; —·—·—, 2.63×10^{-5}; — — —, 3.91×10^{-6} M. The solid line is the absorption in benzene (for Reference) at a concentration of $3.14 \times 10^{-}$ M. (Reprinted with permission from Ref. [176], Copyright 1973, IPC Science and Technology Press)

The electronic spectra of the poly(dienyl)lithiums depend upon the nature of the solvent employed, and to some extent upon the concentration. Poly(isoprenyl)lithium exhibits an absorption maximum in n-hexane that varies from 272 to 275 nm as the concentration is changed from about 10^{-3} to 10^{-6} molar [76]. At the lowest concentrations there is the development of a shoulder centered on 315 to 320 nm (Fig. 13). These results were interpreted in terms of the dissociation of tetrameric aggregates to dimers on dilution. A similar spectral change was noted on changing temperature at constant concentration. Subject to the severe experimental difficulties and the assumptions noted by the authors regarding the interpretation of the spectra, the apparent dissociation constant is much greater in benzene than in n-hexane (viz 10^{-5} and 5×10^{-7} respectively). In contrast, no dependence of the spectrum upon concentration was found for poly(butadienyl)lithium in n-hexane over the range 6×10^{-3} to 2×10^{-5} M.

Somewhat similar spectral changes are observed [191] when ether solutions of $(CH_3)_3CCH_2CH = C(CH_3)CH_2Li$ — a model for poly(isoprenyl)lithium — are

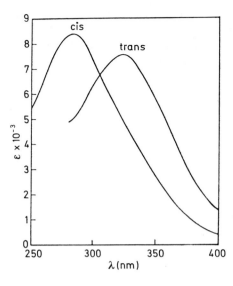

Fig. 14. Spectra of the poly(butadienyl)lithium active center at equilibrium in tetrahydrofuran at various temperatures. (Reprinted with permission from Ref. [192], Copyright 1975, American Chemical Society)

examined. At concentrations in excess of 10^{-3} molar, a shoulder develops at ca. 260 nm (in addition to the principal absorption maximum at 316 nm) which was attributed to the formation of aggregated species.

Garton and Bywater [192] studied the anionic polymerization of butadiene in THF using both lithium and sodium as counterions. They found that the spectrum recorded during polymerization depends upon the temperature as shown in Fig. 14. On completion of the polymerization at -40 °C the spectrum changed, the intensity at longer wavelength decreasing while that at short wavelength increased. A similar sort of change in spectrum is caused by changing temperature. Raising the temperature of the completely polymerized solution caused a similar loss of intensity at short wavelength and a corresponding increase at long wavelength. Analysis of the spectra and comparison with the results of NMR studies showed that the active centers exist in cis and trans forms differing markedly in spectrum (λ_{max} cis 285 nm; λ_{max} trans 325 nm). At equilibrium, the proportion of the cis form increases with decreasing temperature. However, the spectra indicate that the trans conformation is preferentially produced by propagation, the more so the lower the temperature (kinetic and not thermodynamic control). At about 0 °C, the rate of conformational relaxation becomes sufficiently rapid that the spectrum is independent of the occurrence of propagation.

The polymerization of isoprene [193] in THF initiated by poly(styryl)lithium resembles that of butadiene in that there is evidence of the major active center having the cis conformation during polymerization between 0 and -20 °C. Below this temperature the absorption maximum moves to longer wavelength corresponding to the preferential formation of the trans species. The difference between the absorption maxima of the cis and trans ions is quite small, viz. 287 and 305 nm, respectively. Allowing a solution of poly(isoprenyl)lithium in THF to warm to 30 °C results in a chemical change to a species having a broad asymmetric absorption band with a maximum near 330 nm. This process was retarded by the presence of lithium tetraphenylborate — presumably by suppressing dissociation into the free carbanion. Lowering the temperature of such an "isomerized" solution to $-30\ °C$ caused no spectral change. However, if monomer was added to the cooled solution propagation ensued as evidenced by the restoration of the original absorption band at shorter wavelength. The identity of the species absorbing at 330 nm was not established other than that it is not

$$\sim\!\!-\!\!\overset{\displaystyle CH_2}{\underset{\displaystyle CH_2}{\overset{\displaystyle /\!/}{\underset{\backslash\backslash}{C}}}}\!\!Li$$

formed by a rearrangement of the chain end as had been proposed previously [194].

Under favorable circumstances NMR spectroscopy can provide unequivocal evidence regarding the stereochemistry of the anionic end group in poly(dienyl)-lithiums. The immediate problem of detecting signals due to the end group in the presence of in-chain units can be solved either by (a) using oligomers of very low DP or (b) using an "invisible" polymer "capped" by a "visible" end group. In the case of (a) it is assumed that the behavior of an oligomer is identical to that of a polymer — an assumption probably first questioned by Makowski and Lynn [41]. By way of illustration, the 1:1 adduct of butadiene with t-butyllithium is reported as 42% cis at $-18\ °C$ by Glaze et al. [195] in pure THF and as 40% cis by Bywater et al. [139] in equimolar THF/methylcyclohexane at $0\ °C$. The latter authors, however, found that the oligomer (DP = 6) is 66% cis in pure THF at $0\ °C$. They noted also that the γ resonances shift slightly upfield with increasing DP to a limiting position reached by (or before) DP = 6. In the alternative approach (b) the in-chain units may be per-deuterated and the terminal unit hydrogen containing when ^{1}H-NMR spectroscopy is to be employed. Where ^{13}C-NMR is planned, capping may be performed with a ^{13}C enriched monomer.

Morton et al. [135, 141] were the first to study the poly(butadienyl)lithium anionic chain end using (b). They found no evidence of 1,2-chain ends and concluded that only 1,4-structures having the lithium σ-bonded to the terminal carbon were present. A later study by Bywater et al. [196], employing 1,1,3,4-tetradeuterobutadiene to minimize the complexity of the spectrum that arises from proton-proton coupling, found that the 1:1 adduct with d-9 $tert$-butyllithium in benzene exists as a mixture of the cis and trans conformers in the ratio 2.6:1. Glaze et al. [36] obtained a highly resolved spectrum of neopentylallyllithium in toluene and found a cis:trans ratio of about 3:1.

Bywater et al. [139] found that the mole fraction of the trans conformer of poly-(butadienyl)lithium in THF drops from 0.34 at $0\ °C$ to 0.17 at $-40\ °C$; calculation shows that this corresponds to an enthalpy change of ca. 3 kcal/mole for the process trans $\rightarrow$ cis carbanion. Table XII contains data obtained by Glaze et al. [36] on the ^{1}H-NMR spectrum of neopentylallyllithium in benzene, diethyl ether and THF. It is at once evident that there is a pronounced and progressive upfield shift as the polarity of the solvent is increased: the α protons simultaneously move downfield. These changes correspond to the movement of negative charge from the α to the γ position and the transformation of a *substantially covalent* structure into one that is more ionic and delocalized. ^{13}C-NMR spectroscopy of the same solution yielded [196] the data summarized in Table 12. The upfield shift of the γ carbon resonance consequent upon increase of solvent polarity mimics that of the γ proton. An interesting and unexpected observation is that the cis:trans chain end ratio is the same in benzene as it is in ether but that this is very different from that in THF. Clearly there

Table 12. ^{1}H-NMR Data for Neopentylallyllithium [36, 195] (5,5-dimethylhex-2-enyllithium)

Solvent	Temperature	Isomer	δ_α	δ_β	δ_γ	J Hz	$J_{\beta\gamma}$ Hz
Toluene	20	trans	0.77	6.06	4.64	9.4	14.5
		cis	0.80	6.11	4.50	10.0	10.0
Ether	21	trans	1.04	5.97	4.06	10.1	14.3
		cis	1.01	6.08	3.89	10.0	10.5
THF	30	trans	1.29	5.98	3.63	10.5	13.4
		cis	1.19	6.10	3.35	9.9	9.9

Table 13. C-NMR Data for the 1:1 Adduct of t-Butyllithium with Butadiene or Isoprene [36, 196–198]

Monomer	Solvent	Temp.	Isomer	δ_α	δ_β	δ_γ	Counter-ion	% trans-isomer
Butadiene	Benzene	20°	trans	20.7	144.3	103.3	Li	77
			cis	20.0	140.3	103.0		
Butadiene	Ether	−20°	trans		144.1	84.4	Li	75
			cis	30.7	140.0	87.6		
Butadiene	THF	−20°	trans	—	146.0	79.6	Li	35
			cis	31.0	142.5	81.9		
Isoprene	Benzene	20°	trans	27.8	150.6	100.0	Li	65
			cis	24.2	148.4	100.4		
Isoprene	Ether	−20°	trans	36.7	150.9	81.8	Li	25
			cis	33.3	149.2	87.6		
Isoprene	THF	−20°	trans	31.8	149.3	83.0	Li	0
Isoprene	THF	−20°	—	36.7	146.1	79.2	Na	0
Isoprene	THF	−20°	—	45.8	143.2	71.9	K	0
				53.7	143.4	70.8	Cs	

is no simple relationship between charge density and conformational preference. Similar studies have been made on 2,5,5-trimethylhex-2-enyllithium; a model of the propagating center of isoprene [196–198]. A selection of ^{13}C data is collected in Table 13.

There is an even more striking change in the conformational preference of poly-(isoprenyl)lithium on increasing solvent polarity than observed in the poly(butadienyl)-lithium model. There is no detectable trans-carbanion in THF, and even changing solvent from benzene to ether causes a decrease in the trans content from 65 to 25%. In THF, the exclusively cis conformation is noted [198] for the sodium and potassium ion pair of the isoprene model: the butadiene model carbanion is 22% trans with sodium but less than 10% trans with potassium (all at −20 °C).

The influence of the addition of THF to a solution of oligomeric poly-(butadienyl)lithium in benzene has been investigated by Bywater et al. [139]. To simplify the spectrum, the 1,1,3,4-d-4 monomer was employed. With the oligomer having a DP of ca. 6 at 0 °C, the percentage of the cis conformer changed in the sequence 37, 59, 66% as the % THF changed from 10 to 20 to 100%.

Milner and Young [181] have made a similar study of the influence of TMEDA upon the ^{1}H-NMR spectrum of poly(butadienyl)lithium in benzene at 30 °C. In the absence of added base the γ resonance consists of three peaks: the two at lowest field (Fig. 15) correspond to the trans and cis conformers having a penultimate unit enchained in a 1,4-sense, the highest field signal representing carbanions of either conformation preceded by a 1,2-unit. Two signals for the β allylic proton correspond to the cis and trans conformations. The addition of a small amount (r = 0.05) of TMEDA generates a new γ resonance at high field while the center peak of the original trio, most probably due to the cis conformer, diminishes in intensity. Increasing r to 0.4 causes a marked upfield movement of the γ proton signals; two are evident, that showing the greatest shift attributable to the cis chain end the other signal representing the trans. At the same TMEDA level, the β resonances have moved downfield. The three resonances may be identified, moving downfield, as due to chain ends preceded by a 1,2-unit, trans and then cis, both preceded by 1,4-units. It is clear that complexation by TMEDA results in increased ionic character of the chain end. Further additions of TMEDA cause greater upfield shifts of the γ proton signals and downfield shifts of the β, reaching a limit at a level of r in excess of unity. These results are interesting in that (a) one conformer is preferentially complexed by TMEDA and (b) the cis:trans ratio is relatively insensitive to TMEDA. Clearly, (a) may give rise to a rather complex dependence of propagation rate upon r. Comparison of the influence of TMEDA and of THF upon the location of the γ proton resonances underscores the remarkable complexing power of the former agent.

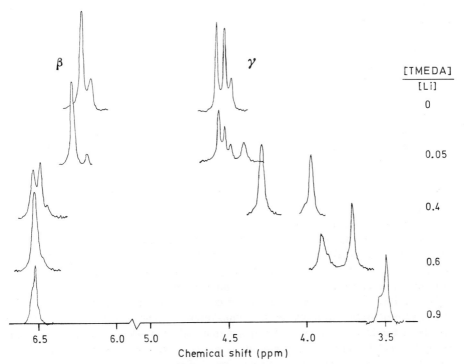

Fig. 15. Influence of the addition of TMEDA upon the spectra of the β and γ protons of poly(α4-butadienyl)lithium. (Reprinted with permission from Ref. [181], Copyright 1982, IPC Science and Technology Press)

By way of illustration, to achieve the spectral shift created by TMEDA at a value of r of 0.5 and 1.0, it is necessary to add ca. 20 and 35 % by volume of THF [139]. With both complexing agents, lowering the temperature increased the fraction of the trans conformer, showing that the equilibrium is mobile although sufficiently slow with respect to the NMR timescale to enable the observation of separate resonances for the two conformers. Dumas et al. [90] have found that the addition of small amounts of TMEDA to poly(isoprenyl)lithium changes the proportion of the cis conformation of the chain end from 40 % in the absence of base to 100 % at a base:lithium ratio of about 0.4:1. Brownstein et al. [197] also report a predominance of the trans conformation for hydrocarbon solutions of the 1:1 adduct of t-butyllithium and isoprene. Morton et al. [142] obtained very similar spectra for oligo(isoprenyl)lithium in benzene but assigned the γ resonances to the two conformers differently than did Brownstein et al. [197]. The former group did not explain their assignment; the latter group did-they argued by analogy with the butadiene model on the basis of the observed coupling constants. There is no evidence of a γ proton resonance from the chain end of oligo-2,3-dimethyl(butadienyl)lithium in benzene, indicating the structure is 1,4-localized and the absence of chain ends corresponding to 1,2-addition [142]. The chain ends derived from penta-1,3-diene in hydrocarbon solvents show some interesting differences from those from isoprene or butadiene [199–201]. Using ethyllithium as the initiator, the γ proton resonances were observed at 4.6δ, consistent with a localized 4,1-structure. However, when isopropyl- or sec-butyllithium was employed there were also γ resonances at 3.1δ. Taken in conjunction with the triplet pattern of the γ signals, it was concluded that a much more delocalized, or ionic, 1,4-chain end was also present. The differences among the initiators was attributed to differences in the relative initiation:propagation ratios; longer chains proportionately less 1,4-structured chain ends. The influence of the presence of large amounts of unconsumed initiator was not determined. Oligohexa-2,4-dienyllithium in benzene also has a ^{1}H-NMR spectrum that is indicative of a more highly delocalized, or ionic, structure [199–201]. The γ proton resonance is at 3.1δ. — approximately the same chemical shift as that for the γ proton of poly(butadienyl)-lithium in THF. Hydrocarbon solutions of poly(pentadienyl)lithium and poly(hexa-dienyl)lithium are distinguished from those of poly(isoprenyl)lithium and poly-(butadienyl)lithium by being strongly colored, i.e., red or yellow.

The ^{13}C-NMR spectra of a model for the propagating chain of butadiene have been determined (Table 14) [139, 196]. Mirroring the behavior of the ^{1}H-spectrum, a change of solvent from benzene to THF causes a large upfield shift for the γ carbon atom, a small upfield shift for the β and a large downfield shift for the α carbon. These changes may be interpreted as arising from conversion to a more highly delocalized carbanion on moving from a nonpolar to a polar solvent. Accompanying this is a change in the balance of the conformational equilibrium in favor of the trans structure. Interestingly, although the chemical shifts in ether and in THF are quite similar, the cis:trans ratios are quite different. The factors that govern the latter are far from clear. In these unsymmetrical ions there is a preference for charge to reside on the α carbon, the more so the smaller the cation. It is possible to make a more quantitative estimate of charge distribution if some plausible assumptions are made. The generation of unit charge on a single carbon usually causes an upfield shift of about 130 ppm. Comparison of the ^{13}C-NMR spectra of propene

Table 14. Calculated charges on Allylic Carbon Atoms of (I) Neopentylallyl[a]- and (II) Neopentyl-methylallyl[b]-Alkali Metal Compounds [139,196]

Metal	Solvent	α	β	γ	Σ[c]	α	α	β	Σ[c]
Li	Benzene	0.79	−0.13	0.22	0.88	0.80	−0.14	0.19	0.85
	Ether	0.69	−0.13	0.35	0.91	0.72	−0.15	0.30	0.87
	THF	0.69	−0.15	0.40	0.94	0.73	−0.15	0.34	0.92
Na	THF	0.65	−0.12	0.49	1.02	0.69	−0.12	0.38	0.95
K	THF	0.59	−0.11	0.53	1.01	0.61	−0.09	0.44	0.96
Rb	THF	0.55	−0.11	0.53	0.97	0.58	−0.09	0.47	0.96
Cs	THF	0.51	−0.12	0.52	0.91	0.54	−0.10	0.45	0.89

[a] $t\text{-}C_4H_9CH_2CH=CHCH_2M$;
[b] $t\text{-}C_4H_9CH_2CH=C(CH_3)CH_2M$;
[c] Σ total charge

and allyllithium in THF leads to the conclusion that the change in hybridization from sp^3 to sp^2 causes a downfield shift of 97 ppm. On this basis, the charge densities of Table 14 were calculated from the observed chemical shifts. Although the absolute accuracy of these calculations may be subject to query, the trend to increasing electron density at the γ position is highlighted.

The structures of benzylic carbanions have been extensively studied and a summary of the more important recent results is presented in Table 15. The ^{1}H-NMR spectrum of benzyl potassium [202] in THF is upfield of that of benzyllithium [203] as expected on the basis of the relative polarizing powers of the two cations. The spectrum of oligomeric poly(styryl)lithium which is qualitatively similar, is shifted upfield on

Table 15. Proton Chemical Shifts for Benzylic Anions [202–205]

Compound	Solvent	ortho	meta	para	α	β	CH$_3$
Benzyllithium	THF	6.09	6.30	5.50	1.62		
Styryllithium[a]	C_6H_6	5.98	6.53	5.52			
Styryllithium	92% C_6H_6 + 8% THF	6.00	6.62	5.32	3.02	2.06	
Styryllithium	THF	5.87	6.32	5.12	2.36	1.85	
Benzyl Potassium	THF	5.59	6.12	4.79	2.24		
Cumyl Potassium	THF	5.15	6.08	4.41			1.48
α-Methylstyryllithium[b]	C_6H_6	5.67 5.29	6.23	4.66		1.60	1.20
α-Methylstyryllithium	THF	5.57 5.19	6.10	4.46		1.88	1.51
Dimer α-methylstyryl potassium	THF	5.38 4.80	6.01 5.89	4.20			1.32
Dimer 1,1-diphenyl-ethylene potassium	THF	7.01	6.55	5.67			2.48

[a] Mean composition $C_4H_9(CH_2CHPh)_2^- Li^+$;
[b] Major component $t\text{-}C_4H_9(CH_2CMePh)^- Li^+$;
[c] $K + PhCMeCH_2CH_2CMePhK^+$;
[d] $K^+ Ph_2CCH_2CH_2CPh_2K^+$

changing the solvent from benzene to THF. Unfortunately, it is not possible to provide a simple interpretation of this change purely on the basis of ionic character since the active chain ends aggregate to form dimers in benzene, but not in THF [42, 56]. The α-methyl(styryl)lithium oligomer is of special interest since the ortho protons (and to a smaller extent the meta-) are inequivalent, showing at once the sp^2 character of the benzylic carbon atom and the existence of a significant barrier to rotation of the phenyl group. Raising the temperature removes this inequivalence. It is surprising that no such inequivalence is shown by either cumyl potassium [203] nor by the dimeric dianion of 1,1-diphenylethylene [204]. The latter species is also unusual in the marked low-field position of its spectrum, which in part is due to non-coplanarity.

^{13}C-NMR spectroscopy of the dimeric dianions of α-methylstyrene and diphenyl ethylene (Table 16) parallels the ^{1}H-NMR spectra in that the ortho carbon atoms of the former ion are inequivalent, in contradistinction to those of the latter [206]. Comparison of the α-methylstyrene dianion spectrum with that of the related neutral hydrocarbon 2,5-diphenylhexane provides particularly clear evidence of the deshielding of the α carbon atom arising from the change from sp^3 to sp^2 hybridization.

Table 16. CMR Chemical Shifts for PhCCH$_2$CH$_2$CPh in THF [205]
$$\text{PhCCH}_2\text{CH}_2\text{CPh} \quad \text{with R, R = K, K}$$

	Ipso	Ortho	Meta	Para	α	β	CH$_3$
R = C$_6$H$_6$	145.8	117.5	129.2	108.0	86.9	30.4	
R = CH$_3$	137.5	103.5	129.6	88	78.4	33.6	19.1
		107.9	131.5				
2,5-diphenylhexane	148.4	127.5	128.9	126.5	41.4	37.4	22.9

10 Stereochemistry of Polydienes

The stereochemistry of the polymerization of dienes is most conveniently discussed in two sections (a) polymerization in hydrocarbon solvents and (b) polymerization in the presence of amines, ethers and other electron donors.

10.1 Polymerization in Hydrocarbon Solvents

The conjugated dienes can polymerize in four modes: cis 1,4-, trans 1,4-, 1,2- and 3,4-, the latter pair being equivalent in the absence of appropriate substitution. Early workers relied entirely upon IR spectroscopy to analyze the concatenation in their polymers. There are a number of problems associated with the technique: correct assignment of peaks, the additivity and the inherent insensitivity arising from the smallness of the extinction coefficients of double bonds bearing more than one substituent (such as arises from 1,4-enchainment). In consequence, the reliability of much of the early work is uncertain; the advant of NMR spectrometers has, however, transformed this situation. An extensive compilation of literature data has been assembled by Yudin [207].

One of the most attractive features of the polymerization of isoprene by organolithium initiators is the high content of the cis 1,4-structure in the product. The figures quoted in the literature vary considerably — probably not only as a consequence of errors arising from analysis, but also from failure to recognize that the microstructure is sensitive to the concentration of initiator, and to a lesser extent that of monomer. Very high cis-1,4-content is only obtained (Table 17) when very low concentrations of organolithium are employed in the absence of solvent while the trans content is enhanced by the use of aromatic solvents and by higher chain end concentrations. The connection between the stereochemistry of the in-chain units and that of the carbanionic centers is not obvious, the latter having apparently exclusively 4,1-structure and with a trans:cis ratio [196] of about 65:35.

Table 17. Microstructure of Polydienes Obtained using Organolithium Initiators

[Initiator] mol l^{-1}	Solvent	Temp. °C	1,4-Trans	1,4-Cis	1,2	3,4	Ref.
Polyisoprene							
6×10^{-3}	Heptane	−10	18	74	—	8	[208]
1×10^{-3}	Heptane	−10	17	78	—	5	[208]
1×10^{-4}	Heptane	−10	11	84	—	5	[208]
8×10^{-6}	Heptane	−10	—	97	—	3	[208]
9×10^{-3}	Benzene	20	25	69	—	6	[209]
4×10^{-5}	Benzene	20	24	70	—	6	[209]
1×10^{-2}	Hexane	20	25	70	—	5	[209]
1×10^{-5}	Hexane	20	11	86	—	3	[209]
3×10^{-3}	None	20	18	77	—	5	[209]
8×10^{-6}	None	—	0	96	—	4	[209]
Polybutadiene							
5×10^{-1}	Benzene	20	62ᵃ		38	—	[181]
5×10^{-2}	Benzene	20	83ᵃ		17	—	[181]
5×10^{-3}	Benzene	20	93ᵃ		7	—	[181]
5×10^{-1}	Cyclohexane	20	53ᵃ		47	—	[139]
5×10^{-2}	Cyclohexane	20	90ᵃ		10	—	[139]
5×10^{-3}	Cyclohexane	20	93ᵃ		7	—	[139]
8×10^{-6}	Benzene	20	52	36	12	—	[209]
1×10^{-5}	Cyclohexane	20	68	28	4	—	[209]
3×10^{-2}	Hexane	20	30	62	8	—	[209]
2×10^{-5}	Hexane	20	56	37	7	—	[209]
3×10^{-3}	None	20	39	52	9	—	[209]
7×10^{-6}	None	20	89	86	5	—	[209]
Poly(1-phenyl)butadiene							
2×10^{-2}	Benzene	20	59	25	—	16	[210]
	Hexane	20	49	28		23	[210]
Poly(penta-1,3-)diene							
—	Hexane	22	40	49	11	—	[211]

ᵃ Total of cis and trans forms

However, the absence of a simple correlation need not be totally surprising if propagation is only through the intermediacy of a minute proportion of chains present in a non-aggregated form with the NMR spectrum reflecting the structure in the predominant aggregates.

Worsfold and Bywater [212] have proposed that propagation through non-aggregated chains is kinetically (and not thermodynamically) controlled and yields only the carbanion having the cis conformation: this can isomerize to the trans form unless the geometry is locked in by a further act of propagation:

$$\sim\sim cis^* \; + \; monomer \longrightarrow \sim\sim cis, cis^*$$
$$\big\updownarrow$$
$$\sim\sim trans^* + \; monomer \longrightarrow \sim\sim trans, cis^*$$

The rate of conformational isomerization in heptane was determined for model compounds I, II and III.

$$\underline{t}\text{-BuCH}_2\text{CH}{=}\overset{\overset{\textstyle CH_3}{|}}{C}\text{CH}_2\text{Li} \qquad \underline{t}\text{-BuCH}_2\text{CH}{=}\overset{\overset{\textstyle CH_3}{|}}{C}\text{CH}_2\text{CH}{=}\overset{\overset{\textstyle CH_3}{|}}{C}\text{CH}_2\text{Li} \qquad \underline{n}\text{-BuCH}_2\text{CH}{=}\overset{\overset{\textstyle CH_3}{|}}{C}\text{CH}_2\text{Li}$$

$$\text{I} \qquad\qquad\qquad \text{II} \qquad\qquad\qquad \text{III}$$

The rate of isomerization of I was found to be given by the equation;
$\text{Rate} = 1.9 \times 10^{13} \exp\left(-\dfrac{10,500}{T}\right) s^{-1}$, which is too slow to account for the observed dependence of polymer stereochemistry upon chain end concentration. However, it was found that species II isomerizes some twenty times faster than I — sufficiently fast to be of significance in this regard. Interestingly, the isomerization of III is faster still. Species II was formed from I by the addition of a molecule of monomer: the kinetic study showed that the initial conformation to be >90% cis. The observation of first order relaxation with a normal pre-exponential factor suggests that relaxation occurs within the aggregates.

That the chain end concentration influences the geometry of the polymer is due to the one-fourth order dependence of propagation upon this quantity. To illustrate this point, consider the effect of increasing chain ends by a factor of 10^4. The propagation rate will increase by a factor of only 10 so that at the higher concentration the chain will, on average, have the interval between successive acts of propagation increased by 10^3, so allowing a greater degree of conformational relaxation to occur.

Morton and Rupert [209] have presented microstructure results for polybutadiene and polyisoprene as a function of conversion and temperature. (Tables 18 and 19).

The results of Table 20 show that the microstructure obtained on polymerizing 2,3-dimethylbutadiene in heptane depends upon the initiator concentration and upon the pressure to which the solution is subjected [213]. At constant pressure, an increase in the concentration of butyllithium results in a decrease of the 1,2-enchainment and an increase in 1,4-trans; the 1,4-cis remains constant. This behavior contrasts with that of isoprene where the vinyl addition is increased on increasing initiator concentration. The geometry of the chain end (0.2 molar) has been reported [142] as entirely 1,4-, but the conformation does not seem to have been established. Increasing

Table 18. Effect of Conversion on Chain Structure of Polyisoprene[a]

Conv. (%)	Trans-1,4	Microstructure[b] Cis-1,4	3,4
13	2	93	5
29	3	92	5
40	2	92	6
46	3	92	5
48	1	95	4
86	1	95	4

[a] No solvent; polymerization temp. = 20 °C; [s-BuLi] = 10^{-5} M;

[b] Via 300 MHz ^{1}H-NMR

Table 19. Effect of Polymerization Temperature on Polydiene Chain Structure

Polyisoprene[a]

Temp. (°C)	[s-BuLi] 10^3	Solvent	Microstructure[b]			
			Trans-1,4	Cis-1,4	3,4	1,2
46	.01	None	1	95	4	—
20	.01	None	1	95	4	—
0	.01	None	2	95	3	—
−25	.01	None	3	93	4	—
40	1	Hexane	18	76	6	—
−25	1	Hexane	18	78	4	—

Polybutadiene[b]

35	.007	None	9	85	—	6
20	.007	None	9	86	—	5
0	.007	None	7	86	—	5

[a] Monomer = 2.5 M;   [b] Via 300 MHz ^{1}H-NMR

Table 20. Influence of Pressure and Initiator Concentration upon the Microstructure of Poly-2,3-dimethylbutadiene [monomer] = 3.75 molar

[BuLi] (Molar)	1,4-Cis %	1,4-Trans %	1,2 %
Pressure-1 bar			
0.03	42	28	30
0.09	35	40	25
0.19	35	56	9
0.25	38	54	8
Pressure-6000 bars			
0.03	47	9	43
0.09	46	15	40
0.18	47	25	28
0.25	47	29	24

Solvent — Heptane

pressure results in increasing fractions of 1,4-cis enchainment. Beilin et al. have suggested [214] that the cis enchainment of butadiene and of isoprene occur on the monomeric form of the chain end while the "formation of the other units is connected with the associates". In the absence of experimental data, particularly relating to the aggregation constants for the poly(dienyl)lithiums, it is difficult to comment upon this proposal. It is worth emphasizing that the assumption of inactivity in aggregates is based solely upon the belief that there is a necessary connection between the degree of aggregation and the kinetic order. The possibility of direct reaction of monomers with the aggregated forms of organolithium active centers has been mentioned previously in this review.

From their study of the ^{1}H-NMR spectra of oligo(dienyl)lithiums prepared from butadiene and isoprene, Morton et al. concluded [141, 142] that in hydrocarbon solution the lithium is essentially σ-bonded to the terminal carbon atom with no detectable 1,2- or 3,4-structure in the carbanion. In order to account for the respective in-chain 1,2- and 3,4-units they concluded that there is an undetected small (less than 1%) content of an exceptionally highly reactive species. Two possibilities were envisaged for isoprene:

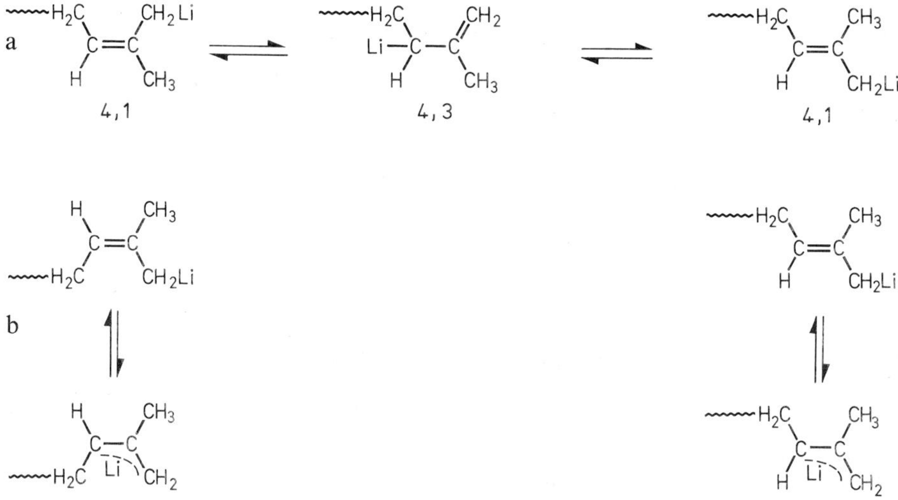

In (a) there is a 4,1 ⇌ 4,3 tautomeric equilibrium in which the proposed minute amount of 4,3-chain end is of sufficient reactivity to account for the in-chain 3,4-units amounting to some 10% of the chains. In alternative (b) a covalent σ ⇌ ionic π equilibrium exists which is heavily in favor of the covalent structure; again the π-form is supposedly highly reactive, leading to in-chain 1,4 and 3,4 placement. A fundamental difference between these proposals is that whereas (a) leads to trans 4,1 ⇌ cis 4,1 carbanion isomerization, proposal (b) does not. Morton et al. [142] found that on storing their samples at 50 °C for a few days, some decomposition took place and the cis:trans ratio of the living ends increased if the original initiator was sec-butyllithium, but decreased if it was iso-propyllithium. This difference in behavior was attributed to the different cross-associates formed with unconsumed initiator which was present in large amounts. This pattern of different alternation of conformation was taken as evidence for (b). However, it is clear that there is facile cis-trans isomerization in poly(butadienyl)lithium [212] and it is most unlikely that this is not also so for poly(isoprenyl)lithium. Accordingly, it would seem that proposal (b) can be discounted, at least in the form originally proposed.

More compelling evidence for a π-allyl, or at least a more highly delocalized chain end, was reported for penta-1,3-diene by Morton et al. [199, 200]. In benzene, using isopropyllithium as initiator, they observed ^{1}H-NMR resonances at 3.1δ due to the γ proton of the 1,4-end and at 4.6δ due to the γ proton of the 4,1-end. The remarkably high field position of the former γ proton (analogous to that of poly-(butadienyl)lithium in THF) corresponding to a large charge density necessarily implies significant delocalization.

An interesting study [215] of the reaction with n-butyllithium in toluene of hydrocarbons (A) and (B), respectively analogous to the cis and trans isomers of non-rigid dienes, has shown that (B), unlike (A), is unreactive:

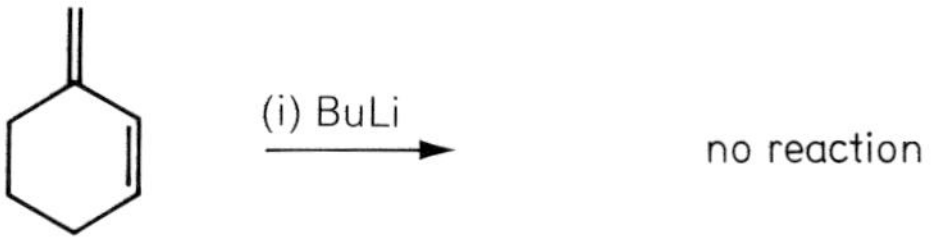

The polymerization of 1-phenylbutadiene by lithium alkyls in hydrocarbon solvents results in 50–60% trans-1,4, some 25% cis-1,4 and 10–25% 3,4-enchainment [210]. When THF is employed as solvent, the corresponding values are about 80, 10 and 10% respectively.

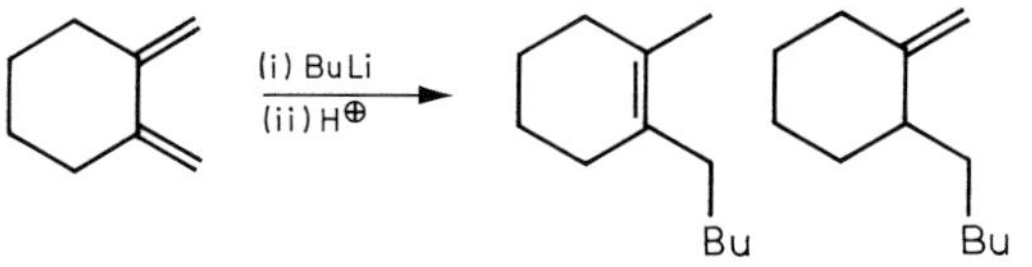

Inomata [211] studied the ^{1}H-NMR spectra of poly(penta-1,3-diene) and concluded that with hexane as polymerization medium the polymers were about 49% cis-1,4 and 40% trans-1,4 enchained. The polymer derived from the cis monomer had 12% of 1,2-units which were exclusively trans; that from the trans monomer had some 10% of 1,2-units, two thirds of which were trans. Aubert et al. [216] made a more extensive study of pentadiene polymers using both ^{1}H and ^{13}C-NMR spectroscopy and modified the cis and trans-1,4 methyl resonance assignments made by Inomata [211].

The stereochemistry of diene polymerization is somewhat dependent upon temperature; note however the results of Morton and Rupert [209] (Table 19). In general, linear Arrhenius plots are obtained. Some pertinent data are summarized in Table 21; for convenience results are presented for both solvating and non-solvating media. The structures obtained in solvating media are very different from those obtained in hydrocarbon solvents (cf Tables 17 and 22). The sensitivities to tempera-

Table 21. Enthalpy and Entropy Differences[a] for Different Addition Modes

Monomer	Solvent	$\Delta_v^b - \Delta_{1,4}$		$\Delta_{1,4cis} - \Delta_{1,4trans}$		Ref.
		ΔH	ΔS	ΔH	ΔS	
Butadiene	Cyclohexane	1.16	0.38	0.90	0.18	[217]
Butadiene	Cyclohexane/THF	−3.67	11.8	−0.11	1.22	[217]
Butadiene	Dioxan	−2.9	−5.5			[218]
Isoprene	Benzene	0.5	4.0	−2.0	−4.0	[219]
Isoprene	Dioxan	−5.7	−14.7			[219]
		−3.3	−8.3	1.9	4.0	[218]
2,3 Dimethylbutadiene	Heptane	−3,8	−4.4			[210]
	Dioxan	−1.2	−4.0			[210]
	THF	−0.98	−3.70			[220]
	THF	−3.25	−12.87			[220]

[a] ΔH in kcal mol^{-1}, ΔS in cal mol^{-1} K^{-1};
[b] Δ_v refers to the sum of 1,2 and 3,4 addition;
[c] above −30 °C;
[d] below −30 °C

Table 22. Microstructure of Polydienes Prepared in Solvating Media

Solvent	Cation	Temp. (°C)	Cis 1,4 %	Trans 1,4 %	3,4 %	1,2 %	Ref.
Butadiene							
THF	Li	80		25		70	[218]
THF	Li	15		13		87	[218]
THF	Na	0	6	14		80	[192]
Dioxan	Na	15		15		85	[218]
	K	15		45		55	[218]
	Cs	15		59		41	[218]
Radical Polymer			25	50		25	[227]
Isoprene							
THF	Li	30		12	59	29	[228]
THF	free anion	30		22	47	31	[228]
THF	free anion	−70		10	45	45	[219]
Ether	Li	20		35[a]	52	13	[191]
Ether	Na	20		17[a]	61	22	[191]
Ether	K	20		38[a]	43	19	[191]
Ether	Cs	20		52[a]	32	16	[191]
Dioxan	Li	15	3	11	68	18	[218]
TMEDA[b]	Li			30	55	15	[229]
TMEDA[c]	Li			25	45	30	[230]
DME	Li	15					[231]
Radical Polymer			63	25	6	6	[227]

[a] cis and trans-1,4 not separated but predominance of trans expected;
[b] Principal solvent was benzene; base: Li ratio was 60:1;
[c] Principal solvent was hexane; base: Li ratio was 1:1

ture are also dissimilar being more marked in solvating media. With isoprene and butadiene vinyl content decreases slightly on lowering the temperature in hydrocarbons but rises quite steeply in ethers. The dependence of the microstructure of poly-2,3-dimethylbutadiene upon reaction conditions shows some interesting features [220]. A high percentage of 1,4-addition occurs in hydrocarbon solvents. In THF at 0 °C there are almost equal amounts of 1,2- and 1,4-addition whereas at −78 °C the 1,2-content is 86%. An Arrhenius plot of ln ([1,4]/[1,2]) vs 1/T exhibits a change of gradient at about −30 °C, corresponding to a higher activation energy at low temperature. It was suggested that the sterically hindered 1,2-placements are interrupted by 1,4-placements ever more frequently as the temperature increases until at −30 °C and above there are alternating placements. These structural observations have been confirmed and elaborated [221].

^{13}C-NMR spectroscopy has shown that the polybutadienes prepared using alkyl-lithium initiators have random placement of the different modes of enchainment [222, 223]. This contrasts with an earlier claim of blocky structures [224]. Random sequence distribution has also been established for polyisoprene by ^{1}H-NMR [225] and ^{13}C-NMR [226] spectroscopy.

10.2 Polymerization in Solvating Media

Under the conditions customarily employed (chain end concentration ca. 10^{-3} molar) for polymerization in solvating solvents such as ethers or amines, there is no significant tendency for aggregation of the propagating centers. However, complications can arise as a consequence of the simultaneous presence of ion pairs and free carbanions. Because of the exceptionally high reactivities of free carbanions and loose ion pairs compared to the modest reactivity of tight ion pairs, the course of the polymerization may be dominated by the presence of very small amounts of the two former species. The contribution of free carbanions can readily be determined by monitoring the consequences of suppressing dissociation by the introduction of a highly dissociated unreactive salt having a common cation. Because of the very small dissociation constants for poly(styryl)lithium, and the even smaller constants for poly(dienyl)-lithiums, a minute quantity of added common cation will completely suppress dissociation.

A selection of data relating the stereochemistry of polyisoprene and polybutadiene to the preparative conditions is shown in Tables 22 and 23. The extent of 1,2-addition of butadiene is very high in comparison with the situation in hydrocarbon solvents and is decreased in favor of trans 1,4-addition by raising the temperature or by using a heavier alkali metal cation in dioxan or THF [192, 218]. When the major component of the solvent is a hydrocarbon and the solvating base is added in modest proportions the resulting stereochemistry is sensitive to temperature to an extent that depends on the base:lithium ratio (Table 23). Weak donors such as ether and triethylamine need to be present in large proportions to engender much 1,2-addition, whereas the powerful chelating agents TMEDA and DPE cause very large percentages of vinyl addition. In general, on raising the temperature of the polymerizing system the influence of the base is reduced. Since the effect of increasing temperature is most marked at low ratios of base:lithium, the phenomenon is most readily ascribed to

Table 23. Dependence of Vinyl Content of Polybutadiene upon
Temperature and Concentration of Base [2, 91]

Base	[Base]/[Li]	% 1,2-addition at		
		30 °C	50 °C	70 °C
Triethylamine	270	37	33	25
Ether	12	22	16	14
Ether	96	36	26	23
THF	5	44	25	20
THF	85	73	49	46
TMEDA	1.14	76	61	46
DPE[a]	1	99	68	31
DPE[a]	10	99	95	84

[a] DPE is 1,2-dipiperidino ethane

increasing dissociation of the lithium-base complex on warming [91, 232]. In accordance
with this interpretation, the sensitivity to temperature is greater the smaller the base:
lithium ratio. The enchainment of butadiene in a virtually 100% vinyl fashion in the
presence of DPE is remarkable [2]. Since similar results are obtained using other bulky
chelating bases, but not with TMEDA, it would seem that steric factors play a
particularly important role in directing the course of propagation. One interesting
phenomenon observed with TMEDA is that considerable cyclization (ca. 60%)
results when the butadiene monomer is introduced into the reactor at low rates [183-185].
Under such conditions propagation does not effectively compete with cyclization;
similar results are obtained with DPE [93].

The polymerization stereochemistry of isoprene is also very sensitive to solvation
or complexation. Addition in a 1,4-sense is generally largely, or exclusively, trans;
addition in the 1,2-mode is accompanied by much larger proportions of 3,4-
enchainment with the only exception [219] being free carbanionic propagation in THF
at −70°. It is worth taking particular note that although ¹H-NMR studies have
only detected the presence of anionic 4,1-chain ends there must, nonetheless, be 1,4-
anions also in order to generate in-chain 1,2-placements:

It has been reported that when employing isoprene enriched with [13]C- at the 1-position, reaction with *t*-butyllithium in benzene resulted in "moderately strong" signals from the 1,4-chain end [233].

Polyisoprene obtained in very powerful solvating media such as DME or hexamethylphosphoramide is largely formed by free anion propagation [231]. Similar polymers are formed in dioxan when the cation is complexed by an appropriate cryptand. In this last case, the interionic separation is presumably sufficiently large that the influence of the cation is minimal [230]. These polymer structures differ most notably from those obtained with ion paired systems in that the latter have a much lower content of 1,2-units. This implies either that a greater fraction of the free anion exists in the 1,4-form than is the case with the ion pairs or, that in the 1,4-chain end the reactivity at the 2-position (relative to the 4-position) is greater in the free ion than in the ion pair. In weakly solvating media having low dielectric constants the only kinetically significant species are the tight ion pairs. Generally, in these solvents increasing cation size results in an increase in the 1,4-trans content at the expense of the 3,4-; the 1,2- is substantially unchanged [191, 218].

Salle and Pham have proposed [220] that in a solvent such as dioxan, the tightly ion paired species poly(isoprenyl)lithium propagates through the intermediary of a monomer-complexed species. Coordination by isoprene involves only the 3,4-double bond due to the relative inaccessibility of the lithium arising from its coordinating dioxan molecules. Propagation yields the trans allylic anion; of a second act of propagation takes place sufficiently rapidly, the trans active end will become an in-chain 1,4-trans or 3,4-unit. If, however, there is a sufficient delay between successive acts of propagation, the anionic chain end can relax to the more stable cis conformation and propagation will thereby result in 1,4-cis or 4,3-addition. This proposed competition between kinetic and thermodynamic control is analogous to that suggested by Worsfold and Bywater [212] for the propagation of poly(isoprenyl)lithium in hydrocarbon solvents.

The tacticity of anionically prepared polystyrenes has been the subject of extensive study by a number of groups of workers, mostly by means of ^{13}C-NMR spectroscopy. From a study of the aromatic Cl resonances, Matsuzaki and coworkers found [234] that there is a tendency towards syndiotacticity when using *n*-butyllithium in toluene as initiator. From the sensitivity of the CMR spectrum to the nature of the solvent employed it was concluded that the polymerization did not conform to Bernoullian statistics. Randall examined the methylene resonances in the CMR spectrum and concluded that butyllithium initiated polystyrene is essentially atactic [235] and that propagation is Bernouillian. Uryu et al. [236] examined polystyrene formed with other alkali metal initiators in several solvents. They found that in THF and DME with Na, K, Rb and Cs counterions, the values of P_r fall in the range 65 to 69%. In diethyl ether at 30 °C, they found a shift towards atacticity with increasing cation radius: Na (66%), K (60%), Rb (53%) and Cs (55%). At lower temperature (−78 °C) with the heavier cations, a slight bias towards isotacticity was evident: Rb (46%) and Cs (44%). It must be said, however, that there are some disquieting features in their preparative work. Fluorenylcesium was used as an initiator, although it is generally claimed to be unreactive in this capacity. Furthermore, the description of the reaction mixture of styrene and sodium naphthalene in THF as having a "greenish red" color is alarming and the interpretation that this reflects incomplete consumption of initiator despite being accompanied by the generation of high molecular weight polymer is certainly incorrect. Suparno and coworkers [237] obtained values of P_r in THF which are in good agreement with those of Matsuzaki et al. [234]; in toluene, hoever, they found a bias towards syndiotacticity (with K^+ and C_s^+) in contrast to the Japanese group who reported [236] an isotactic bias.

Wicke and Elgert [238] have concluded that the tight ion pair, the loose ion pair and the free carbanion of poly-α-methylstyrene generate the same tacticity. With n-butyllithium as the initiator in THF, the activation energy for meso propagation is a little greater than that for racemic propagation (15.6 and 14.9 kcal/mole respectively); the pre-exponential factors are the same [239].

11 Chain Propagation in Ethers

Perhaps the most noteworthy feature of many anionic polymerizations conducted in ethereal solvents is that the propagation reaction is a dual process, i.e., ion pairs and free ions coexist and exhibit distinctly different reactives [240−244]. The free ion concentration is generally limited to 1% or less of the total active center concentration while the relative reactivities diminish in the sequence free anion ≥ loose (solvent-separated) ion pair ≫ tight (contact) ion pair.

The propagation reaction in these systems is characterized by an expression of the type [240−245]:

$$k_{obs} = k_t c_t + k_l c_l + k_- K_{diss}^{1/2} C_0^{-1/2} \tag{47}$$

where c_t and c_l are respectively the mole fractions of ion pairs present in the tight and loose forms, k_t, k_l and k_- are the rate constants for the tight ion pairs, the loose ion pairs and the free carbanions while C_0 denotes the total active center concentration.

The poly(styryl)lithium active center was found [179] to partially dissociate into the free ion in benzene-THF solution when the mole fraction of THF in the solvent mixtures was > ca. 50%. Solvents or solvent mixtures of lesser polarity generally do not lead to the formation of significant concentrations of the highly reactive free ions, i.e., the ion pair reaction appears to dominate.

Morton and co-workers [126] were apparently the first to study the polymerization of butadiene and isoprene in THF. Their work was subsequently followed by that of Arest-Yakubovich and Medvedev [245]. These combined results indicated that for the lithium counter-ion, the reactions are first order in active centers. Thus, the ion pairs were implicitly assumed to represent the only active species.

However, the later work of Bywater and Worsfold [228] clearly showed that the free-ions do contribute, albeit to a relatively small extent, to the propagation process for isoprene in THF. At 30 °C, the rate constants for the ion pair and free ion were found [228], respectively, to be 0.20 $M^{-1} s^{-1}$ and $2.8 \times 10^3 M^{-1} s^{-1}$ while the dissociation constant was 5.0×10^{-10} M. In diethyl ether at 20 °C, the ion-pair seems, though, to be the sole active center [246]. At -20 °C, the ion pair propagation constant in THF is decreased [193] to ca. $7 \times 10^{-3} M^{-1} s^{-1}$.

The respective propagation constants for isoprene in THF should be considered to be composite values since the polymer prepared in THF contains both 3,4 and 1,2 units in a ratio of about 7/3 [228]. Thus the two active centers

$$-CH_2-HC \underset{Li}{\overset{\overset{\displaystyle CH_3}{|}}{\underset{}{C}}} CH_2 \text{ and } -CH_2-\underset{Li}{\overset{\overset{\displaystyle H_3C \ H}{| \ |}}{\underset{}{C}}} CH_2$$

and their respective free ions will contribute to the polymerization.

Vinogradova et al. [247] have measured the electrical conductivity of poly(butadienyl)-lithium over a wide range of concentration in dimethoxyethane and THF. The former solvent led to an active center dissociation of about 1%. The rate of propagation of poly(butadienyl)lithium was determined [248] in THF and it was concluded that tight ion pairs and free carbanions participate according to the usual law shown in Eq. (47). The plots of k_{obs} vs. $C_0^{-1/2}$ were linear although the authors [248] did not comment on their negative intercept, i.e., negative values of the ion pair rate constant, $k_{\pm}$. At 20 °C for poly(butadienyl)lithium in THF K_{diss} was found to be $5.1 \times 10^{-10} M^{-1}$ while k_- was $2.0 \times 10^3 M^{-1} s^{-1}$. These values are similar to those measured for poly(isoprenyl)lithium under the same conditions. Although Vinogradova and co-workers [248] did not report the ion pair propagation constant for poly(butadienyl)-lithium in THF (20 °C), the value of ca. 0.38 $M^{-1} s^{-1}$ is avalable from the work of Garton and Bywater [249].

12 Copolymerization Involving Diene and Styrenes

Contrary to what has been observed for radical systems, lithium based anionic copolymerizations usually exhibit pronounced sensitivity to solvent type. Thus, the polarity and solvating power of the solvent will influence the copolymer reactivity ratios while mixtures of e.g. ethers and hydrocarbons will lead to effects intermediate with regard to what is observed for the pure solvents.

It has been found (Table 24) that in hydrocarbon solvents, the diene polymerizes preferentially in diene-styrene systems, initially to the near exclusion of the styrene even though the latter monomer exhibits the faster rate of homo-polymerization. When the diene supply nears depletion, styrene begins to become incorporated into

Table 24. Copolymerizations in Hydrocarbon Solvents

M_1	M_2	Solvent	T°C	r_1	r_2	Ref.
1,3-Butadiene	2-Methyl-1,3-butadiene	Hexane	50	3.38	0.47	[250]
			40	1.85[a]	0.30[a]	[251]
			30	1.61[a]	0.35[a]	[251]
			20	2.66[a]	0.40[a]	[251]
		Benzene	40	3.6	0.5	[252]
1,3-Butadiene	Styrene	Benzene	29	4.5	0.08–0.41	[253]
			30	10	0.035	[254]
			30–50	20	0.05	[255]
		Toluene	25	12.5	0.1	[256]
		Cyclohexane	40	26	<0.04	[257]
		Hexane	50	15.1	0.025	[258]
		Heptane	30	7	0.1	[259]
2-Methyl-1,3-butadiene	Styrene	Benzene	30	7.7	0.13	[260]
			29	4.9–23.1	0.13–0.46	[254]
			30	7.0	0.14	[261]
		Toluene	27	9.5	0.25	[262]
		Cyclohexane	40	16.6	0.046	[263]
2-isohex-3-enyl 1,3-butadiene[b]	Styrene	Benzene	30	5.55	0.08	[264]
Styrene	p-tert-Butyl styrene	Benzene	20	1.30	0.84	[265]
Styrene	p-Methylstyrene	Benzene	20	0.72	1.09	[265]
			30	2.5[c]	0.4[c]	[266]

[a] Average values from results determined by five different methods;
[b] Myrcene;
[c] Active center concentration ca. 50 times larger than that ($\sim 10^{-3}$) of Ref. 265

the chain in a significant fashion, and the polymerization rate approaches that of styrene alone.

A parallel situation is encountered for the copolymerization of 1,3-butadiene with isoprene. McGrath et al. [251] have shown that in homopolymerizations, under equivalent conditions, isoprene exhibits a rate constant which is more than five times larger than that observed for butadiene. However, butadiene is favored in the copolymerization. The available reactivity ratios for various diene and styrenyl monomer pairs in hydrocarbon solvents are listed in Table 24.

McGrath's results [251] regarding the rate constants for 1,3-butadiene and isoprene (butadiene < isoprene) place in clear perspective the bizarre [267] assertion that 1,3-butadiene will be twice as reactive as isoprene since" one of the ends of isoprene" relative to butadiene — "is unreactive while the other retains its normal reactivity".

The reversal of reactivity of styrene and the dienes in copolymerizations has been explained on a kinetic basis [253, 263, 268], i.e., that the rate constants for the four possible reactions decrease in the sequence:

$$k_{SD} \gg k_{SS} > k_{DD} > k_{DS}$$

where the subscripts S and D denote styrene and diene respectively and k the appropriate apparent rate constant, the first subscript defining the origin of the active center.

An alternative explanation by Korotkov [254] is available and involves complexation of the active centers by the diene involved. This involves diene solvation of the active center (or centers for the associated species) where it was assumed that styrene is less effective in solvation. Since the advance of this concept by Korotkov [254] in 1958, subsequent evidence, which has been noted previously in this review, has accrued which indicates that aggregated carbon-lithium species can indeed form such complexes via interaction with π-electrons. Obviously, though, the role, if any, of monomer "solvation" in these copolymerizations remains to be elucidated in detail.

An additional point to be clarified in these copolymerizations relates to the question as to whether the self- and cross-aggregated active centers are reactive entities in their own right or merely serve as dormant reservoirs providing unassociated, reactive centers. As has been noted previously in this review a body of results has appeared which involves associated organolithiums as reactive species in either initiation or propagation.

The presence of cross-associated species needs to be considered in the interpretation of copolymerization kinetics. It has been found [269] that the reaction of poly(butadie-nyl)lithium with p-divinylbenzene in benzene solution proceeds at a rate which increases markedly with time. Such a result implies that the poly(butadienyl)lithium aggregate is less reactive than the mixed aggregate formed between the butadienyl- and vinylbenzyllithium active centers. Interestingly, no accelerations with increasing reaction time were found with poly(butadienyl)lithium and m-divinylbenzene nor with poly(isoprenyl)lithium and either the m- or p-divinylbenzenes. This general behavior was subsequently verified [270] by a series of size exclusion chromatography measurements on polydiene stars (linked via divinylbenzene) as a function of conversion.

When an excess of styrene (S) was added to the $\sim\sim CH_2C(C_6H_5)_2Li$ active center, DLi, the ensuing crossover reaction followed pseudo first order kinetics [271];

$$-d[DLi]/dt = k_u[S_0][DLi] . \tag{48}$$

However, a series of such experiments showed that k_u is not constant but is inversely proportional to the square root of the total organolithium concentration. This behavior can be contrasted to that of isoprene adding to styryllithium and styrene adding to isoprenyllithium in cyclohexane [263] where the apparent rate constants *were* constant.

During the crossover reaction of styrene to the DLi active centers, the following equilibria are maintained for the dimeric aggregates:

$$(DLi)_2 \quad \rightleftharpoons 2\,DLi \qquad K_a \tag{49}$$

$$(SLi)_2 \quad \rightleftharpoons 2\,SLi \qquad K_b \tag{50}$$

$$(SLi \cdot DLi) \rightleftharpoons SLi + DLi \qquad K_{ab} \tag{51}$$

The variation in k_u with total active center concentration has been rationalized utilizing the following assumptions:

a) the crossover reaction proceeds solely through the intermediacy of the unassociated chain end, DLi;

b) the concentrations of the unassociated chains are small in comparison with those of the dimeric species; and

c) K_{ab} has a geometric mean relationship to K_a and K_b where:

$$-d(DLi)/dt = k_x[S_0] [DLi] [DLi]_0^{1/2} K_a^{1/2} \qquad (52)$$

where k_x is the rate constant for the addition of styrene to the DLi active center and $[DLi]_0$ denotes the total concentration of active centers.

The reactions of poly(styryl)lithium in benzene with an excess of diphenylethylene [272] and bis[4-(1-phenylethenyl)phenyl]ether [158] also were found to proceed by a first order process. However, the reactions of poly(styryl)lithium with the "double diphenylethylenes" 1,4-bis(1-phenylethenyl)benzene and 4,4′ bis(1-phenylethenyl)1,1 biphenyl gave [158] non-linear first order plots with the gradients decreasing with time. This curvature was attributed to departure from a geometric mean relationship between the three dimerization equilibrium constants (K_a, K_b and K_{ab}). The respective concentrations of the various unassociated, self-associated and cross-associated aggregates involved in the systems described by Equations (49) to (51) are dependent upon the relative concentrations of the two active centers and the respective rate constants which govern the association-dissociation events.

The rate of reaction of disubstituted 1,1′diphenylethylenes with lithium, potassium and cesium polystyryl were found by Busson and van Beylen [272] to yield linear Hammett plots corresponding to ϱ values of 1.8, 2.2 and 2.4 respectively, in benzene at 24 °C. The value of ϱ for the reactions involving poly(styryl)lithium in cyclohexane was also 1.8. The active center concentration range studied was small. These authors interpreted their results on the basis of the equation:

$$-d[DPE]/dt = k_p K_{diss}^{1/2}[DPE] [S^-M^+]_0^{1/2} \qquad (53)$$

by operating under conditions where $[DPE] > [S^-M^+]$. The potential influence of cross-association was not considered; contrast this result with Eq. (52).

The addition of small amounts of a polar solvent can markedly alter the copolymerization behavior of, for example, the diene-styrene pair. The solvation of the active centers manifests itself in two ways: the incorporation of styrene is enhanced and the modes of diene addition other than 1,4 are increased [264, 273]. Even a relatively weak Lewis base such as diphenyl ether will bring about these dual changes in anionic copolymerizations, as the work of Aggarwal and co-workers has shown [260]. Alterations in polyisoprene microstructure and the extent of styrene incorporation were found for ether concentrations as low as 6 vol. % (φ-o-φ/RLi = 54) at which concentration diphenyl ether has been shown [52] to cause partial dissociation of the poly(styryl)lithium dimers. The findings of Aggarwal and co-workers [260] are a clear demonstration that even at relatively low concentrations diphenyl ether does interact with these anionic centers and further serve to invalidate the repetitive claim [78, 158, 160, 161] that diphenyl ether — at an ether/active center ratio of 150 — does not interact with carbon-lithium active centers.

13 Chain Transfer in Anionic Polymerization

The process of chain transfer has received very little quantitative study insofar as the anionic systems are concerned. The first study of an anionic chain transfer process was that of Robertson and Marion [274] on the polymerization of 1,3-butadiene by sodium in toluene. The reaction of toluene with the sodium active center led to the formation of benzyl sodium. This work was the first to demonstrate the important role of solvent in transfer reactions involving anionic active centers:

$$2 \, Na + nCH_2 = CH—CH_2 = CH_2 \rightarrow Na^+ {}^-[C_4H_6]_n^- \, Na^+ \qquad (54)$$

$$Na^+ {}^-[C_4H_6]_n^- \, Na^+ + 2\emptyset CH_3 \rightarrow H[C_4H_6]_n \, H + 2\emptyset CH_2 Na \qquad (55)$$

The ability of toluene to serve as a transfer agent was further demonstrated by Bower and McCormick [275] and Brooks [276] for the organosodium initiated polymerization of styrene in that solvent. Both groups reported molecular weights lower than the values calculated from the monomer-initiator ratio.

Higginson and Wooding [277] also reported a transfer reaction to solvent for the case of the polymerization styrene in ammonia initiated by potassium amide. There was no termination event in their kinetic scheme, i.e., active center deactivation via a spontaneous termination event was not considered to be a significant event.

Gatzke [278] has investigated the chain transfer process involving toluene and poly(styryl)lithium at 60 °C. A relationship between the number-average degree of polymerization and the transfer constant was derived:

$$x_n = [M]X/[PSLi] — C_{RH}[RH] \ln (1 — X) \qquad (56)$$

where [M] denotes the monomer concentration, X the extent of conversion, [PSLi] the active center concentration, [RH] the concentration of toluene. The transfer constant C_{RH} was found to be 5×10^{-6}.

The transfer reaction of butadiene with the sodium counterion in a tetrahydrofuran-toluene solution was studied [279, 280]. The presence of the ether was found to enhance the transfer reaction to the point where a transfer constant near unity was obtained.

Low molecular weight polybutadienes of various mixed microstructures are prepared [281] commercially via an anionic chain transfer process. These polymerizations use toluene as the solvent and transfer agent and lithium as the counter ion. The transfer reactions is promoted by the use of diamines, e.g., tetramethylethylenediamine, or potassium t-butoxide. The preparation, modification, and applications of these materials has been described by Luxton [281].

14 Active Center Stability

The topic of the stability of anionic centers in hydrocarbon solvents was apparently first addressed by Ziegler and Gellert [282] in 1950 for ethyl- and n-butyllithium. n-Butyllithium was found to decompose at temperatures above 100 °C to yield

1-butene ($\sim 92\%$), butane (8%) and lithium hydride. At ambient temperature, though, n-butyllithium is stable. Such, though, is not the case with the branched butyllithium isomers. s-Butyllithium decomposes at a rate of ca. 0.1% active lithium per day at room temperature [283]. Glaze, Lin and Felton [118] examined the thermal decompositions products from s-butyllithium. Lithium hydride and the three isomers of butene were found. Bryce-Smith [284] thermally decomposed t-butyllithium in refluxing heptane and found an isobutylene/isobutane (94/6) mixture. Finnegan and Kutta [285] proposed that lithium hydride is generated via a concerted four-center type transition state for the case of n-butyllithium.

Ethereal solvents react directly with alkyllithiums via either proton abstraction or ether cleavage [4]. Thus, in polar solvents such as ethers, alkyllithiums have, at best, limited stability at room temperature.

Antkowiak [286] and Nentwig and Sinn [287] have studied the thermolytically induced reactions for poly(butadienyl)- and poly(isoprenyl)lithium in hydrocarbon solvents. Their combined findings are shown in the following equations.

$$2\sim CH_2-CH=CH-CH_2Li \xrightarrow{\Delta} \sim CH_2-CH \overset{CH}{\underset{Li}{\diagup \diagdown}} CHLi + \sim CH_2-CH=CH-CH_3 \tag{57}$$

$$\sim CH_2-CH \overset{CH}{\underset{Li}{\diagup \diagdown}} CHLi + D_2O \longrightarrow \sim CH_2-CH=CH-CHD_2$$

$$\text{and } \sim CH_2-CHD-CH=CHD \tag{58}$$

$$\sim CH_2-CH=CH-CH_2Li \xrightarrow{\Delta} \sim CH=CH-CH=CH_2 + LiH \tag{59}$$

$$\sim CH=CH-CH=CH_2 + \sim CH_2-CH=CH-CH_2Li \longrightarrow$$

$$\sim CHLi-CH=CH-CH_2-CH_2-CH=CH-CH_2\sim \tag{60}$$

These results show that the elimination of lithium hydride can occur and that dilithiated chain ends can be formed. The formation of such species may account for the dark red to brown coloration these solutions develop upon heating with no monomer present. Furthermore, the formation of the 'macrodiene' of Equation (59) provides the reactive site which can lead to chain coupling and linking. Ultracentifugation and size exclusion chromatography measurements have shown [287] that three-armed stars can be formed. The presence of these coupling and linking events will obviously distort an initial near-monodisperse molecular weight distribution.

Apparently, reactions similar to those outlined in Equations (57) to (60) occur in dilute ($\sim 10^{-3}$ M) solutions of the delocalized active center based on 2,4-hexadiene [288]. This conclusion is based on both spectral results and analysis by the application of size exclusion chromatography. These reactions, though, are supressed [288] at higher concentrations ($>10^{-2}$ M) of active centers. The transformations which occur under dilute conditions account for the different association numbers [48, 199] (i.e. 1.7 and 1.4) which have been reported. This has been verified by the observation [288] that N_w for freshly prepared hexa(dienyl)lithium active centers (formed by adding

2,4-hexadiene to poly(isoprenyl)lithium in benzene) is ~ 1.7 and that the association state decreases to ~ 1.4 when the solution is held for a period of ca. two weeks at 30 °C.

The active center α-methyl(styryl)lithium undergoes a slow decomposition in benzene at 30 °C. Using ethyllithium as the initiator, Margerison and Nyss demonstrated [289] that the main products of this decomposition process are lithium hydride and 1,3-dimethyl-3-phenyl-1-propylindane. Their reaction sequence follows:

$$(61)$$

As Margerison and Nyss noted [289] this reaction path is similar to one given by Benkeser et al. [290] to explain the formation of 1,1,3-trimethyl-3-phenylindane in the course of the metallation of isopropylbenzene by n-pentylsodium:

$$(62)$$

$$(63)$$

Cyclization is not the only method by which the styryl active centers can decompose. Poly(styryl)sodium can apparently decompose by β-elimination of a hydride ion in the following fashion:

$$\text{~~~CH}_2\text{CHNa} \longrightarrow \text{~~~CH}{=}\text{CH} + \text{NaH}$$

$$(64)$$

$$(65)$$

The existence of the 1,3-diphenylallyl ion was inferred via spectroscopy [291, 292]. Although it was claimed [291] that this decomposition reaction (misnamed by the authors as an isomerization process) was faster when the polymer molecular weight was low, no explanation for this unique effect was provided.

In contrast to the claim of Szwarc and co-workers [291] that the α-methyl(styryl)-sodium active center is stable, measurements from various sources have shown [293–301] that transformations readily occur. The dimer structure formed from α-methyl styrene and sodium is as follows:

$$Na-\underset{\underset{C_6H_5}{|}}{\overset{\overset{CH_3}{|}}{C}}-CH_2-CH_2-\underset{\underset{C_6H_5}{|}}{\overset{\overset{CH_3}{|}}{C}} Na$$

with the tetramer being formed by the addition of monomer in the conventional head-to tail fashion [302–304]. This structure for the tetramer represents a correction of the erroneous structure proposed by Szwarc and co-workers [305–307].

The appearance of a new absorption peak near 430 nm was attributed by Decker et al. [299] and Schmitt [296] to the formation of a 2-phenylallyl carbanion:

$$\sim\!\!\sim\!\!CH_2-\underset{\underset{C_6H_5}{|}}{\overset{\overset{CH_3}{|}}{C}} Na \longrightarrow \sim\!\!\sim\!\!CH=\underset{\underset{C_6H_5}{|}}{\overset{\overset{CH_3}{|}}{C}} + NaH \qquad (66)$$

$$\sim\!\!\sim\!\!CH_2-\underset{\underset{C_6H_5}{|}}{\overset{\overset{CH_3}{|}}{C}} Na + CH_2=\underset{\underset{C_6H_5}{|}}{\overset{\overset{CH_3}{|}}{C}} \longrightarrow \sim\!\!\sim\!\!CH_2-\underset{\underset{C_6H_5}{|}}{\overset{\overset{CH_3}{|}}{CH}} + CH_2=\underset{\underset{C_6H_5}{|}}{\overset{\overset{CH_2Na}{|}}{C}}$$

$$(67)$$

Whilst these reactions may well occur, it is certain that the absorption band of the 2-phenylallyl ion will lie at very much shorter wave length. A similar objection can be levelled at the analogue:

$$\sim\!\!\sim\!\!\underset{\underset{C_6H_5}{|}}{\overset{\overset{CH_3}{|}}{C}}-CH\!-\!\!\underset{\underset{C_6H_5}{}}{\overset{\overset{Li}{}}{C}}\!\!-\!\!-\!\!CH_2$$

to which Ades, Fontanille and Leonard [303] assigned a λ max. of 424 nm.

A finding of importance in regard to the stability of the α-methylstyryl anion is that of Comyn and Glasse [309, 310] that a photochemical process plays a role in the transformation. Their proposed mechanism is as follows:

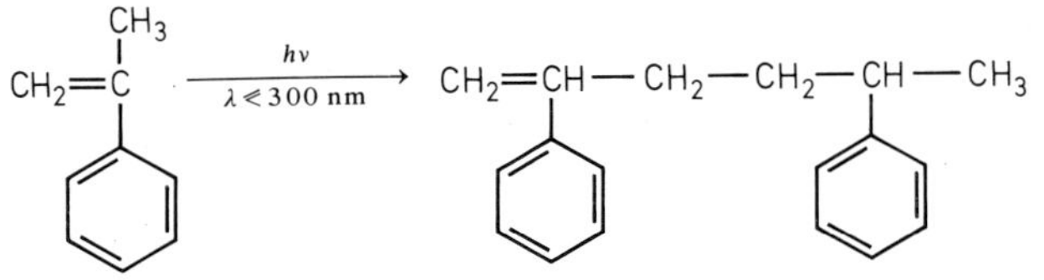

$$
CH_2{=}\underset{\underset{C_6H_5}{|}}{\overset{\overset{CH_3}{|}}{C}} \xrightarrow[\lambda < 300\ nm]{hv} CH_2{=}CH{-}CH_2{-}CH_2{-}CH{-}CH_3 \tag{68}
$$

where the active center serves as a photosensitiser.

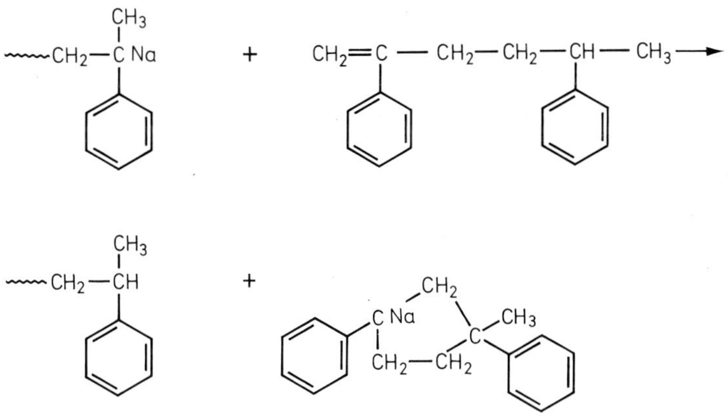

$$
\sim\!\!\sim CH_2{-}\underset{\underset{C_6H_5}{|}}{\overset{\overset{CH_3}{|}}{C}}\ Na \quad + \quad CH_2{=}C{-}CH_2{-}CH_2{-}CH{-}CH_3 \longrightarrow
$$

$$
\tag{69}
$$

$$
\sim\!\!\sim CH_2{-}\overset{\overset{CH_3}{|}}{CH} \quad + \quad
$$

The elimination of the sodium hydride was explained by the process given by Margerison and Nyss [289]. Following Schmitt [296], Comyn and Glasse also proposed [309] that reaction of the anions formed in the α-methylstyrene system would yield deactivated species via reaction with the solvent, THF. Their kinetic study showed [310] that the process given in Eq. (68) was second order in monomer and first order in active centers, which are not consumed in the reaction. The sequence shown as Eq. (69) was found to be first order in active center concentration and in the dimer; which is the product of Eq. (68).

Comyn and Glasse [309] have also suggested that cyclization is a prominent event in the transformation of the styrylsodium active center. This conclusion is based on the work of Margerison and Nyss [289], Benkeser and co-workers [290], and Burley and Young [311].

The foregoing transformation and termination reactions have been studied for the cases where the counter ion was either sodium or potassium. However, there is little doubt that similar reactions can occur involving lithium. There is evidence [312] which suggests that the stability of styryl active centers in ethers is counter ion dependent and changes in the order of Li > Na > K.

The active centers based on styrene, 1,3-butadiene or isoprene and the lithium counter ion in hydrocarbon solvents possess good stability at ambient temperatures over the duration of polymerization and beyond. However, dienyllithium species in ethereal solvents show at best only short term stability [313]. The isomerization processes, involving the cis and trans forms in which these active centers exist, have already been mentioned in this review.

15 Chain end Functionalization

One of the most useful and important characteristics of anionic polymerization is the generation of polymer chains with stable carbanionic chain ends. In principle, these reactive anionic end groups can be readily converted into a diverse array of functional end groups. These end groups may then undergo a variety of further reactions e.g. (1) chain extension, branching or crosslinking reactions with polyfunctional reagents; (2) coupling and linking with reactive groups on other oligomer or polymer chains; (3) initiation of polymerization of other monomers.

The repertoire of reactions possible with organolithium compounds is well documented in the literature [4]. The application of these functionalization reactions to polymers is also described in the anionic polymer review literature [314–316]. Unfortunately, many of the reported applications of these functionalization reactions to anionic polymers have not been well characterized. Accordingly, one is faced with the situation in which a variety of useful chain end functionalization reactions is potentially possible, but whose application to polymers is not well defined in terms of specifics such as side reactions, yields, solvent effects, etc. The following discussion of representative functionalization reactions is not meant to be exhaustive, but can be regarded as typical of the state-of-the-art in this area.

15.1 Carbonation

The carbonation of polymeric anions using carbon dioxide is one of the most useful and widely used functionalization reactions. However, there are special problems associated with the carbonation of polymeric organolithium compounds [317]. For example, Wyman, Allen and Altares [318] reported that the carbonation of poly-(styryl)lithium in benzene with gaseous carbon dioxide produced only a 60% yield of carboxylic acid; the acid was contaminated with significant amounts of the corresponding ketone(dimer) and tertiary alcohol(trimer) as shown in Eq. (70).

$$PSLi \xrightarrow[2) H^+]{1) CO_2(g)} PSCO_2H + (PS)_2CO + (PS)_3COH \qquad (70)$$
$$\quad\quad\quad\quad\quad 60\% \quad\quad\quad 28\% \quad\quad\quad 12\%$$

They also concluded that pouring the active polymer solution onto solid, granulated carbon dioxide produced a total of only 22% of the side reaction products (ketone and tertiary alcohol). A recent paper by Mansson [319] confirmed the effect of quenching with granulated, solid carbon dioxide with reported yields of greater than 90% for the carboxylic acid. In addition it was reported that conversion of poly(styryl)-lithium to poly(styryl)magnesium bromide with magnesium bromide produced the carboxylic acid in high yield (90%) after treatment with gaseous carbon dioxide. Mansson also concluded that "the ability of THF to dissociate dimeric into monomeric species has no dramatic influence on the yield of carboxylic acid", based on the results shown in Eq. (71).

$$PSLi + CO_2(g) \xrightarrow[99.5/0.5]{\text{methylcyclohexane / THF}} PSCO_2H + (PS)_2CO + (PS)_3COH$$
$$\phantom{PSLi + CO_2(g) \xrightarrow[99.5/0.5]{\text{methylcyclohexane / THF}}} \quad 36\% \quad\quad 19\% \quad\quad 38\%$$

(71)

A recent, careful, detailed investigation of the carbonation reaction of polymeric organolithium compounds has revealed several important aspects of this reaction [320]. Using high vacuum techniques and high-purity, gaseous carbon dioxide it has been reported that carbonation of poly(styryl)lithium, poly(isoprenyl)lithium, and poly-(styrene-b-isoprenyl)lithium in benzene produces carboxylic acid in about 60% yield and the corresponding ketone (dimer) in about 40% yield. It is important to note that no tertiary alcohol was reported. It was concluded that tertiary alcohol formation is a side reaction [321] resulting from ketone generation (most probably from water contamination during quenching) in the presence of polymeric organolithium chain ends as shown in Eq. (72).

$$PSLi + CO_2 \longrightarrow PSCO_2Li \xrightarrow{PSLi} PS-\overset{\overset{\displaystyle OLi}{|}}{\underset{\underset{\displaystyle OLi}{|}}{C}}-PS \xrightarrow{H_2O} PS\overset{\overset{\displaystyle O}{\|}}{C}PS \xrightarrow{PSLi} (PS)_3COLi \xrightarrow{H_2O} (PS)_3COH$$

(72)

Contrary to the conclusion of Mansson [319], it would be expected that association of the organolithium chain ends would promote coupling to form the ketone (dimer) product, Eq. (73); conversely, dissociation

$$(PSLi)_2 + CO_2 \longrightarrow (PSCO_2Li)(PSLi) \longrightarrow PS\overset{\overset{\displaystyle OLi}{|}}{\underset{\underset{\displaystyle OLi}{|}}{C}}PS$$
$$ \text{associated}$$

(73)

of the chain ends would be expected to favor formation of the carboxylic acid functionalized chains.

As discussed elsewhere in this review, Lewis bases such as tetrahydrofuran are known to promote disaggregation of polymeric organolithium species [42, 47]. Thus, in the presence of excess tetrahydrofuran, both poly(styryl)lithium and poly(isoprenyl)lithium would be expected to be unassociated (or at least much less associated). Therefore, in the presence of sufficient tetrahydrofuran, the carbonation reaction would take place with unassociated organolithium chain ends and ketone formation (Eq. (73)) would only be an intermolecular reaction (rather than an essentially intramolecular reaction as in the case with the aggregated species) competing with carbonation. In complete accord with these predictions, it was found that the carbonation of poly(styryl)lithium, poly(isoprenyl)lithium, and poly(styrene-b-isoprenyl)lithium in a 75/25 mixture (by volume) of benzene and tetrahydrofuran occurs quantitatively to produce the corresponding carboxylic acid chain ends. The observation by Mansson [319] that THF has no apparent influence was complicated by the use of methylcyclohexane, which is a Theta solvent for poly(styrene) (60–70 °C) [322]; furthermore, the amount of tetrahydrofuran used was probably not sufficient to effect the complete dissociation of the polymeric organolithium aggregates [42].

The carbonation reaction is somewhat ideal since it is possible to analyze the reaction products using a variety of probes including osmometry, size exclusion chromatography end group titration, and thin layer chromatography. Obviously the ability to use some of these analytical methods will decrease with increasing polymer molecular weight. However, many applications of functionalized polymers do not require high molecular weight products.

The carbonation of dilithium reagents is complicated by the occurrence of gelation phenomena which produce severe mixing problems [145, 146, 323]. In general, lithium derivatives of heteroatoms are highly associated in solution; therefore, heteroatom functionalization of polymers with two active anionic chain ends will form an insoluble, three-dimensional network. The beneficial effect of decreasing the effects of association and gelation by solvents with solubility parameters <7.2 has been reported in the literature [140, 324].

With regard to the carbonation of polymeric anions with counter ions other than lithium, Pannell [325] has reported that poly(styryl)potassium reacts with carbon dioxide in tetrahydrofuran to form carboxyl-terminated polymer without the complicating side reactions which generate higher molecular weight species.

15.2 Halogenation

The introduction of halogen end groups on polymers is of considerable interest since organo halogen groups undergo a variety of nucleophilic substitution and elimination reactions as well as serving as potential initiation sites for cationic polymerization. The direct bromination of polymeric organolithium compounds is complicated by competing Wurtz-coupling reactions to yield dimeric products. For example, addition of a benzene THF (250/3) solution of poly(styryl)lithium to an excess of bromine in benzene produced 42% of the coupled product and 58% uncoupled (and presumably)brominated polystyrene (Eq. (74)) [326, 327].

$$\text{PSLi} \xrightarrow[\text{benzene}]{\text{Br}_2} \underset{58\%}{\text{PSBr}} + \underset{42\%}{(\text{PS})_2} \tag{74}$$

It is possible that solvent and Lewis bases could have a significant effect on the amount of coupling product observed, although these variables have not been examined. The lithium chain end has been converted to the corresponding Grignard reagent (PSMgBr) in an attempt to reduce the amount of coupling in the bromination reaction. Thus, reaction of poly(styryl)lithium with a saturated solution of magnesium bromide in tetrahydrofuran followed by quenching with bromine decreased the amount of coupling to 7% and produced 93% of the bromine-terminated polymer (note, however, that no analytical data were presented to document that the uncoupled polymer had one bromine per chain end) [326].

Another route to bromine terminated polymers has been to react anionic polymers with halogenated terminating agents such as α,α'-dibromoxylene [326, 327]. It is presumed that the prevailing reaction is Wurtz-coupling, (Eq. (75));

$$PSLi + \quad\quad\quad \longrightarrow \quad PSCH_2- \quad\quad\quad + LiBr \qquad (75)$$

although simple lithium-halogen exchange could be occurring in this system also, (Eq. (76)).

$$PSLi + \quad\quad\quad \longrightarrow \quad PSBr + \quad\quad\quad \qquad (76)$$

Coupling to produce dimeric product was a side reaction in these systems also, e.g. 75% dimer formation was reported for poly(styryl)lithium and 23% dimer formation with the poly(styryl)-Grignard reagent [326]. However, it should be noted that the only reported characterizations of these reactions were size exclusion chromatography traces and silver catalyzed polymerization of tetrahydrofuran using the polymeric halogen compounds as co-initiator.

In view of the importance of preparing well characterized, halogen terminated polymers, there is an obvious need for a careful examination of the direct halogenation reaction. Optimization of the halogenation reaction, however, may not be straight-forward, since it has been observed by ESR spectroscopy that radicals are formed in the reaction of simple alkyllithiums with bromine or iodine in the presence of equimolar amounts of Lewis bases such as N,N,N',N'-tetramethylethylenediamine or ether [328].

A potential alternative to direct halogenation utilizes agents such as ethylene dibromide. For example, it has been reported that neophyllithium in ether reacts at room temperature with ethylene dibromide to yield neophylbromide quantitatively (Eq. (77) [329]).

$$\begin{array}{c} CH_3 \\ | \\ C_6H_5-C-CH_2Li \\ | \\ CH_3 \end{array} + BrCH_2CH_2Br \longrightarrow \begin{array}{c} CH_3 \\ | \\ C_6H_5-C-CH_2Br \\ | \\ CH_3 \end{array} + CH_2{=}CH_2 + LiBr \qquad (77)$$

One cannot simply extrapolate results obtained for simple alkyllithiums to polymeric organolithiums since important factors such as the degree of association and diffusion rates are different. Thus, preliminary examination of the reaction of poly(butadienyl)-lithium in benzene with excess ethylene dibromide in benzene produced predominately the dimeric coupling product (Eq. (78) [330]).

$$PBdLi + BrCH_2CH_2Br \longrightarrow \underset{24\%}{PBdBr} + \underset{76\%}{(PBd)_2} \qquad (78)$$

This type of halogenation procedure involving active centers should be carefully examined since like α,α'-dibromoxylene it is applicable to unsaturated polymeric anions such as poly(butadienyl)lithium and poly(isoprenyl)lithium whose double bonds would react directly with halogens.

It has been reported that a tetrahydrofuran solution of the disodium salt of the tetramer of α-methylstyrene reacts with iodine to form coupled products with molecular weights of up to 3×10^3 [331]. Although it would be expected that iodine would favor the coupling reaction [4], the lack of specific product and yield information precludes further discussion of this result.

15.3 Ethylene Oxide and Ethylene Sulfide Termination

In contrast to many other functionalization reactions, termination of living anionic polymers with ethylene oxide, (Eq. (79)) is relatively free of side reactions other than polymerization. For example,

$$PLi + CH_2\overset{O}{\diagup}\overset{}{\diagdown}CH_2 \xrightarrow[H\oplus]{} PCH_2CH_2OH \tag{79}$$

Reed [332] has reported that reaction of ethylene oxide with the α,ω-dilithiumpolybutadiene in predominantly hydrocarbon media (some residual ether from the dilithium initiator preparation was present) produced telechelic polybutadienes with hydroxyl functionalities (determined by infrared spectroscopy) of 2.0 ± 0.1 in most cases. A recent report by Morton, et al. [146] confirms the efficiency of the ethylene oxide termination reaction for α,ω-dilithiumpolyisoprene; functionalities of 1.9_9, 1.9_2 and 2.0_1 were reported (determined by titration using Method B of ASTM method E222-66). It should be noted, however, that term of α,ω-dilithiumpolymers with ethylene oxide resulted in gel formation which required 1–4 days for completion. In general, epoxides are not polymerized by lithium bases [333, 334], presumably because of the unreactivity of the strongly associated lithium alkoxides [64] which are formed. With counter ions such as sodium or potassium, reaction of the polymeric anions with ethylene oxide will effect polymerization to form block copolymers (Eq. (80) [334–336]).

$$P^\ominus M^\oplus + nCH_2\overset{O}{\diagup}\overset{}{\diagdown}CH_2 \xrightarrow[H\oplus]{} P(CH_2CH_2O)_{n-1}CH_2CH_2OH \tag{80}$$

Hydroxyl-terminated polymers have also been prepared [337, 338] using organolithium initiators with protected hydroxyl functionality. Thus, using initiators such as 2-(6-lithio-n-hexoxy)tetrahydropyran (A) and ethyl 6-lithiohexyl acetaldehyde acetal (B), it was

$$LiCH_2(CH_2)_5O\text{—}\underset{O}{\diagdown}\diagup$$
$$A$$

$$LiCH_2(CH_2)_5\underset{\underset{OCHCH_3}{|}}{\overset{OCH_2CH_3}{}}$$
$$B$$

possible to prepare narrow molecular weight distribution polybutadiene polymers with either one (0.87–1.02) or two (1.76–2.04) hydroxyl functionalities per chain after mild acid hydrolysis of the acetal groups [337]. The difunctional chains were formed by either terminating with ethylene oxide or coupling with dichlorodimethylsilane. These initiators are insoluble in hexane, but are reported to be soluble in benzene and diethyl ether. A summary of these results as well as a discussion of related patent literature are contained in a recent review [338].

The reaction of α,ω-dilithiumpolyisoprene with ethylene sulfide crosses over very rapidly (4–5 min) at –40 °C in a 50/50 hydrocarbon/tetrahydrofuran mixture, while the subsequent polymerization requires several days [339]. These results suggest that it might be possible to prepare ethane thiol terminated polymers under certain conditions.

15.4 Amination

Several new methods for preparing amine terminated polymers have been described recently. One of the major challenges has been to synthesize polymers with primary amine functionality since primary amine hydrogens undergo proton transfer to anionic chain ends [277]. Schulz and Halasa [340] have prepared —NH$_2$ terminated polydienes using the initiator p-lithio-N,N-bis(trimethylsilyl)aniline (C), which has a primary amine protecting group. Using this initiator, relatively

$$[(CH_3)_3Si]_2N-\!\!\!\bigcirc\!\!\!-Li$$

C

narrow molecular weight distribution (1.06–1.25) polybutadiene was prepared with 69–100% amination as determined by titration after acid hydrolysis of the amine protecting group (Eqn. (81, (82)).

$$[(CH_3)_3Si]_2N-\!\!\!\bigcirc\!\!\!-Li + \underline{n}\ CH_2\!=\!CH\!-\!CH\!=\!CH_2 \xrightarrow[CH_3OH]{} [(CH_3)_3Si]_2-N-\!\!\!\bigcirc\!\!\!-\!\!\left[BD\right]_n\!-H \quad (81)$$

$$[(CH_3)_3Si]_2N-\!\!\!\bigcirc\!\!\!-\!\!\left[BD\right]_n\!-H \xrightarrow{H_3O^{\oplus}} H_2N-\!\!\!\bigcirc\!\!\!-\!\!\left[BD\right]_n\!-H \quad (82)$$

Termination of these polymerizations with dichlorodimethylsilane followed by hydrolysis of the protecting group generated polymeric diamines with functionalities of 1.7–1.9 and relatively broad MW distributions (1.49–2.22). The authors considered that the titrimetric method was less reliable for the higher molecular weight polymers and perhaps is a reason for the apparent ineficiency of this amination procedure. A major limitation of this method is the fact that the initiator is insoluble in hydrocarbon solvents and therefore most of the diene polymerizations were carried out in mixtures of hexane and ether which has a deleterious effect on the microstructure of the diene polymers.

Hirao, et al., [341]) have described a very useful method for the amination of living anionic polymers. Polymeric anions were reacted with a trimethylsilyl derivative of an aldimine (D) which generated the primary amine-terminated polymers after quenching with dilute acid (Eq. (83)).

$$PSLi + \underset{\underset{H}{\overset{H_5C_6}{\diagup}}}{\overset{\diagdown}{C}}=N-Si(CH_3)_3 \xrightarrow[H_3O^\oplus]{} PS-\overset{\overset{C_6H_5}{|}}{C}H-NH_2 \tag{83}$$

$$\underset{D}{}$$

Amine-terminated polymers were obtained in yields of 90–100% for poly(styryl)-lithium ($\bar{M}_n = 1 \times 10^4$ to 1.9×10^5) and poly(isoprenyl)lithium ($\bar{M}_n = 1.74 \times 10^4$). It is noteworthy that using the corresponding sodium and potassium derivatives of polystyrene resulted in decreased yields of aminated polymers (48% and 17%, respectively). These workers also examined the efficiency of other aminating agents (E–G), but all of these reagents gave lower amination yields than the corresponding aldimine derivative (D).

$$BrCH_2CH_2N[Si(CH_3)_3]_2 \qquad \underset{\underset{H_5C_2}{\overset{H_5C_6}{\diagup}}}{\overset{\diagdown}{C}}=N-Si(CH_3)_3 \qquad \underset{\underset{H_9C_4}{\overset{H_5C_6}{\diagup}}}{\overset{\diagdown}{C}}=N-Si(CH_3)_3$$

$$\qquad E \qquad\qquad\qquad F \qquad\qquad\qquad G$$

The analogous reaction of α,ω-disodiumpolystyrene with benzylideneaniline has been reported to proceed in good yield to give the corresponding secondary amine end groups (Eqn. (84)) [342].

$$Na^{\oplus\ominus}[\underset{\overset{|}{C_6H_5}}{C}HCH_2\!\!-\!\!(PS)\!\!-\!\!CH_2\!-\!\underset{\overset{|}{C_6H_5}}{C}H]^\ominus Na^\oplus + 2\,C_6H_5N\!\!=\!\!CHC_6H_5 \xrightarrow{CH_3OH}$$

$$\tag{84}$$

$$C_6H_5NHCH\!-\!\underset{\overset{|}{C_6H_5}}{C}H\!-\!\underset{\overset{|}{C_6H_5}}{C}H_2\!\!-\!\!(PS)\!\!-\!\!CH_2CH\!-\!CHNHC_6H_5$$

Another potentially useful amination procedure utilizes the reaction of organo-lithium compounds with mixtures of methoxyamine and methyllithium (Eqn. (85)) [343]; for example:

$$RLi \xrightarrow[\text{2) } H_2O]{\text{1) } CH_3ONH_2 / CH_3Li} RNH_2 \tag{85}$$

Although benzyllithium has been aminated to 97% yield using these reagents [343] attempted amination of poly(styryl)lithium ($\bar{M}_n = 2 \times 10^3$) was achieved with only 5% efficiency using literature procedures [344]. After extensive modification of those procedures, poly(styryl)lithium has been aminated with 92% efficiency using a two-fold excess of methoxyamine/methyllithium [344]. In addition, pure 1° amine-terminated polystyrene can be isolated by silica gel chromatography since it is easily separated from the unaminated polymer.

Various attempts have been made to utilize lithium amide derivatives as initiators for polymerization of vinyl monomers. Because of the high degree of association of heteroatom lithium derivatives, reagents such as lithium morpholinide [345,346], lithium diethyl amide [347], and other lithium dialkylamides [347] are insoluble in hydrocarbon media. Thus, although these initiators are capable of generating poly-dienes with high 1,4-enchainment in hydrocarbon media, the ability to precisely control molecular weight and molecular weight distribution is lost. However, even in diethyl ether where lithium diethyl amide is soluble, there is a complicating induction period and no simple relationship exists between rate and initiator concentration [347]. From a synthetic point of view, the amine end group functionality of the polymers prepared from these initiators has generally not been well characterized.

Eisenbach, Schnecko and Kern [348] have prepared 3-dimethylaminopropyllithium as an initiator for anionic polymerizations. This initiator generated tertiary amine groups in polymers from α-methylstyrene, and butadiene [349]. The molecular weights of poly-(α-methylstyrene) calculated from the nitrogen content were significantly higher than those determined by vapor phase osmometry; all of the observed molecular weights were much higher than expected from the initial ratio of monomer to initiator.

15.5 Termination with Oxygen and Sulfur

The reaction of active polymer chain ends with oxygen is a complex reaction which can lead to a variety of products depending on the reaction conditions. This reaction is interesting because it offers the possibility of generating a macromolecular peroxide initiator which could be used to form new block copolymers. One of the most common observations from oxygen termination is the presence of coupling products of double the molecular weight of the initial living polymer; alcohol and peroxide functionality have also been detected [350]. Brossas and co-workers [350] have carried out a detailed study of the influence of reaction conditions and anionic chain end-group on the course of the oxidation of poly(styryl)lithium. For example, 10% versus 20% coupling was observed for the inverse addition mode as compared to the direct oxidation mode in THF at −65 °C. The corresponding hydroperoxide functionality (see below) was obtained in 95% yield, while the reverse method gave a 40% yield. Thus, the hydroperoxide can be prepared in high yield by the low temperature oxidation in THF by adding the living polymer to an oxygen-saturated THF solution (inverse addition) (Eq. (86)).

$$\text{PSLi} + O_2 \xrightarrow[\text{inverse addition}]{-78\ °C} \underset{9\%}{\text{dimer}} + \underset{91\%}{\text{PSO}_2\text{H}} \tag{86}$$

Similar results, i.e., efficient formation of the hydroperoxide, were observed by oxidation of the living chain ends in the solid state. As noted by Fetters and Firer [351] from air termination reactions, more coupling product is observed in the presence of Lewis bases such as THF and N,N,N′,N′-tetramethylethylenediamine. These results are consistent with oxidation being the primary reaction responsible for dimer formation in air termination [352].

A simplified radical chain mechanism for oxidation of polymeric organolithiums can be deduced from studies of the oxidation of simple alkylithiums (Scheme 1) [53], assuming the intermediacy of free radicals[1]. One often neglected facet of these oxidation reactions

Scheme 1

$$(PLi)_n + O_2 \rightarrow P \cdot + (P_{n-1}Li_n)^+ + O_2^- \cdot$$

$$P \cdot + O_2 \rightarrow POO \cdot$$

$$POO \cdot + (PLi)_n \rightarrow (POO)(P_{n-1}Li_n) + P \cdot$$

is termination (Scheme 2), as shown for poly(styryl)lithium.

Scheme 2

$$2PSO_2 \cdot \rightarrow [PSO_4PS] \rightarrow (PS)CH_2\overset{\overset{\displaystyle O}{\parallel}}{C}C_6H_5 + (PS)CH_2CH(OH)C_6H_5$$

$$PSO_2 \cdot + PS \cdot \rightarrow PSO_2PS$$

$$2PS \cdot \xrightarrow{\text{combination}} (PS)_2$$

$$2PS \cdot \xrightarrow{\text{disproportionation}} (PS)CH=CHC_6H_5 + (PS)CH_2CH_2C_6H_5$$

The products shown for decomposition of the tetroxide of polystyrene are deduced from studies of the analogous decomposition of 1-phenylethylperoxy radicals [353]. Other reactions which should be considered in these oxidations include hydrogen transfer from solvent (Eq. (87)) [350] and reaction of the

$$P \cdot + RH \rightarrow P-H + R \cdot \tag{87}$$

organolithium compound with the lithium peroxide product (Eq. (88)) [354]

$$PLi + PO_2Li \rightarrow 2 POLi \tag{88}$$

With this background, the following range of products could be forme in the oxidation of a polymeric organolithium compound (Eq. (89)) as illustrated for polystyrene.

$$PSLi \xrightarrow[2)\,H_2O]{1)\,O_2} PSH + (PS)_2 + PSO_2PS + PSO_2H + (PS)CH_2\overset{\overset{\displaystyle O}{\parallel}}{C}C_6H_5 +$$
$$+ PSCH=CHC_6H_5 + (PS)CH_2CH(OH)C_6H_5 \tag{89}$$

Any rigorous study of the oxidation of polymeric organolithium compounds should consider these products and their variation in yield with reaction conditions. To date, few of these reaction products have been considered, let alone identified and analyzed. However, the presence of the macroperoxide has been identified recently among the products of the oxidation of poly(styryl)lithium [352]. Lithium aluminium hydride reduction followed by SEC analysis of the dimer fraction before and after reduction

---

1 It is noteworthy that although evidence for free radical intermediates was observed for oxidation reactions in hydrocarbon solution, a pathway not involving free radicals was proposed for oxidation reactions in the presence of Lewis bases [53]

was used to determine the amount of macroperoxide product formed (Eq. (90)):

$$(PS)_2 + PSO_2PS \xrightarrow[\text{2) } H_3O^+]{\text{1) } LiAlH_4} (PS)_2 + 2PSOH \tag{90}$$

Using the normal addition procedure (O_2 diffusion into a 75/25 benzene/THF solution of poly(styryl)lithium) the 37% "dimer fraction" analyzed for 19% alkyl radical dimer and 18% macroperoxide after $LiAlH_4$ reduction. The yield of macroperoxide was also confirmed by thermal decomposition experiments in refluxing toluene, followed again by size exclusion chromatography analysis of the dimer fraction. The amount of hydroperoxide could be deduced from the difference between the amounts of total peroxide (determined by iodometric titration) *versus* the amount of macroperoxide determined by $LiAlH_4$ reduction.

Mansson [319] has also investigated the products of the reaction of poly(styryl)-lithium with oxygen. Although products with carbonyl and alcohol functionality were detected, they may have resulted from the column and thin layer chromatographic work-up procedures employed. In conclusion, the oxidation of polymeric organo-lithium compounds is complex, but the possibility of manipulating the reaction conditions to form useful macroperoxides and hydroperoxides is real, as evidenced by the work of Brossas and coworkers [350].

It may be possible to prepare lithium thiolate derivatives of living anionic polymers using higher ratios of lithium to S_8, since *sec*-butylthiolate was obtained in *ca.* 60% yield from *sec*-butyllithium in benzene when the ratio was 8 [350].

$$PLi + S_8 \rightarrow P[S]_{3-4}P \tag{91}$$

Brossas and coworkers [350] have reported that reactions of oligomeric organo-lithium compounds with elemental sulfur (S_8) in benzene ($PLi/S_8 = 2$) yield primarily coupled products (*ca.* 80%) with 3–4 sulfur atoms in the bridge (Eq. (91)).

15.6 Miscellaneous Reactions

Sulfinate and sulfone functional groups have been formed [355] by the reaction of sulfur dioxide with partially lithiated polystyrene. This reaction scheme follows:

$$(92)$$

The degree of functionalization was found to be *ca.* 32%; a consequence of the fact that the lithiation reaction itself led to metallation of about 32% of the available aromatic groups. These findings indicate that carbanionic chain ends would also react in an effective fashion with sulfur dioxide.

Following Milkovich [356], Masson, Franta and Rempp [357] have prepared polystyrene terminated with an ester group, i.e.,

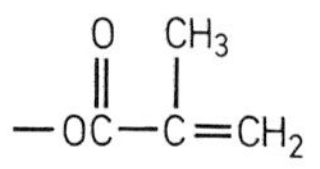

$$-O\overset{\overset{\displaystyle O}{\|}}{C}-\overset{\overset{\displaystyle CH_3}{|}}{C}=CH_2$$

The synthesis of this material was accomplished as follows:

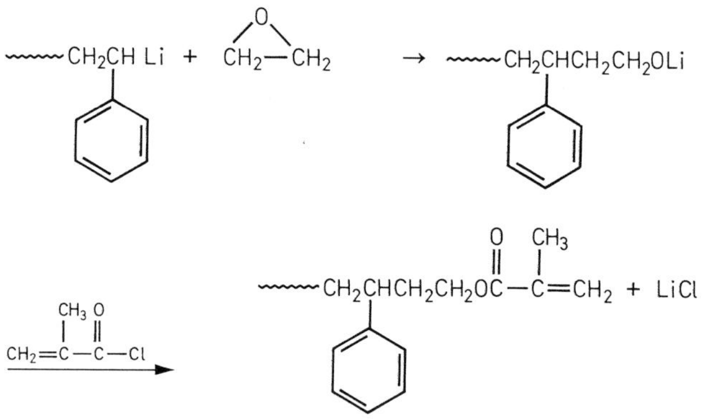

(93)

The addition of ethylene oxide serves to reduce the active center reactivity leading to a reduction in potential side reactions which could occur concurrently with the functionalization step. Masson and coworkers [357] found that the above reaction exhibited an efficiency of about 80%. This can be compared with a parallel attempt to react vinyl benzylchloride to the alkoxylithium chain ends. It was found [358], contrary to Milkovich's claim [356], that the reaction in question did not take place.

Schulz and Milkovich [359] have recently presented a detailed description of their results on the use of functionalized polymers to prepare graft copolymers. They also described the conversion of poly(styryl)lithium to the corresponding lithium alkoxide derivative with ethylene oxide, followed by reaction with methacryloyl chloride (Eq. (93)). This methacrylate-terminated polystyrene was then copolymerized with various vinyl monomers under free radical conditions to form graft copolymers. The functionalized polystyrenes were described as having "very high monofunctionality," although the chain end functionality was not independently characterized. Only size exclusion chromatography analyses of the results of the grafting reactions were presented to verify this description of the efficiency of the functionalization reaction.

16 Appendix

The viscometric technique has been used to determine equilibrium constants for systems such as:

$$(RM_jLi)_2 + 2nE \rightleftharpoons 2(RM_jLi \cdot nE) \tag{a}$$

where:

$$K = \frac{(RM_jLi \cdot nE)^2}{(RM_jLi)_2(E)^{2n}} \tag{b}$$

and RM_jLi denotes, for example, poly(styryl)lithium and E a complexing agent such as an ether. This method relies on the inherent sensitivity of the flow behavior of concentrated polymer solutions in the entanglement regime.

The concentrated solution viscosity measurement yields the weight-average degree of association of active chain ends rather than the more conventional number-average (mole fraction) value. However, the calculation of the equilibrium constant for association, K, can be accomplished if N_w and the heterogeneity index of the polymer sample are known. The latter parameter can be determined via post-polymerization characterization.

The calculation of the equilibrium constant is as follows. Equation (a) can be written as:

$$\frac{C_0}{2}(1 - \alpha) + 2n(E_0 - C_0\alpha) \rightleftharpoons \alpha C_0 \tag{c}$$

where: C_0 = total RM_jLi concentration;
E_0 = total ether concentration; and
α = the fraction of C_0 complexed with the ether.
This permits Eq. (c) to be recast as:

$$K = \frac{2\alpha^2 C_0}{(1 - \alpha)(E)^{2n}} \tag{d}$$

The number-average degree of association can be expressed as:

$$N_n = \frac{2}{1 + \alpha}$$

where

$$\alpha = \left(\frac{2}{N_n}\right) - 1 \tag{e}$$

The substitution of (e) into (d) yields

$$K = \frac{(2 - N_n)^2 C_0}{N_n(N_n - 1)(E)^{2n}} \tag{f}$$

Equation (f) can be converted to an expression involving N_w. Prior to this, though, the relationship between N_n and N_w must be developed. A mixture of associated and unassociated chains will have a bimodal molecular weight distribution ($N_w > N_n$). The following development assumes that the bimodal distribution consists of two monodisperse fractions. Obviously, this is not strictly true, but the fact that the anionic polymerization systems involved can yield a near-monodisperse collection of chains means that little error is introduced by this assumption.

The weight-average association number can be expressed as:

$$N_w = \frac{^a\Sigma\, n_i M_i^2}{^t\Sigma\, n_i M_i^2} = \frac{^a n_1 + {}^a n_2 4}{^a n_1 + {}^a n_2 2} \tag{g}$$

where a and t denote the active and terminated chains and n_1 and n_2 the number of active complexed chains and dimeric species respectively. Putting $x = n_1/n_2$, we then have

$$N_w = \frac{1 + 4x}{1 + 2x} \tag{h}$$

or $\qquad x = \dfrac{N_w - 1}{2(2 - N_w)} \qquad\qquad\qquad$ (i)

Equation (i) has been derived previously [47]. In a similar fashion, since

$$N_n = \frac{^a\Sigma n_1 M_i}{^t\Sigma n_i M_i} \tag{j}$$

then: $\qquad N_n = \dfrac{1 + 2x}{1 + x} \qquad\qquad\qquad$ (k)

The substitution of Eq. (k) into (i) yields:

$$N_n = \frac{2}{3 - N_w} \tag{l}$$

The substitution of Eq. (l) into (f) yields:

$$K = \frac{2(2 - N_w)^2\, C_0}{(N_w - 1)\,(E)^{2n}} \tag{m}$$

Equations (f) and (m) can be used to calculate the equilibrium constant thru the use of either N_n or N_w. For samples with near-monodisperse molecular weight distribution both Eqs. (f) and (m) yield the same value for the equilibrium constant. This situation does not hold though, for polymer systems having polydisperse molecular weight distributions. Thus, the equilibrium constant results presented by Van Beylen and co-workers [360, 361] for polydisperse polystyrenes having the barium or strontium counter ion were weight-average values, not the required number-average based values.

17 References

1. Ziegler, K., Dersch, F., Walltham, H.: Ann., *511*, 13, 45, 64 (1934)
2. Halasa, A. F., Lohr, D. F., Hall, J. E.: J. Polym. Sci., Polym. Chem. Ed., *19*, 1357 (1981)
3. Coates, G. E., Green, M. L. H., Wade, M.: *Organometallic Compounds*, Vol. 1, "The Main Group Elements," Methuen, London, 3rd ed., 1967
4. Wakefield, B. L.: *The Chemistry of Organolithium Compounds*, Pergamon Press, Oxford, 1974
5. Brown, T. L.: Advances Organomet. Chem., *3*, 365 (1965)
6. Brown, T. L.: Pure Appl. Chem., *23*, 447 (1970)
7. Brown, T. L.: Accts. Chem. Res., *1*, 23 (1968)
8. Dietrich, H.: Acta. Cryst., *16*, 681 (1963); J. Organomet. Chem., *205*, 291 (1981)
9. Weiss, E., and Hencken, G.: J. Organomet. Chem., *21*, 265 (1970)
10. Zerger, R., Rhine, W., Stucky, G.: J. Am. Chem. Soc., *96*, 6048 (1974)
11. Rundle, R. E.: J. Phys. Chem., *61*, 45 (1957)
12. Sanderson, R. T.: *Chemical Periodicity*, Reinhold, New York, 1960
13. Schoenberg, E., Marsh, H. A., Walters, S. J., and Saltman, W. M.: Rubber Rev., *52*, 526 (1979)
14. Richards, D. H.: Chem. Soc. Rev., *6*, 235 (1977)
15. Streitwieser, Jr., A., Williams, Jr., J. E., Alexandratos, S., and McKelvey, J. M.: J. Am. Chem. Soc., *98*, 4778 (1976)
16. Streitwieser, Jr., A.: J. Organomet. Chem., *156*, 1 (1978)
17. Graham, G. D., Marynick, D. S., Lipscomb, W. N.: J. Am. Chem. Soc., *102*, 4572 (1980)
18. Graham, G., Richtsmeier, S., and Dixon, D. A.: J. Am. Chem. Soc., *102*, 5759 (1980)
19. Cowley, A. H., and White, W. D.: J. Amer. Chem. Soc., *91*, 34 (1969)
20. Peyton, G. R., and Glaze, W. H.: Theor. Chim. Acta., *13*, 259 (1969)
21. Hein, F., and Schram, H.: Z. Phys. Chem., *A151*, 234 (1930)
22. Brown, T. L., and Rogers, M. T.: J. Am. Chem. Soc., *79*, 1859 (1957)
23. Brown, T. L., Dickerhoff, D. W., and Baffus, D. A.: J. Am. Chem. Soc., *84*, 1371 (1962)
24. Brown, T. L., Gerteis, R. L., Baffus, D. A., and Ladd, J. A.: J. Am. Chem. Soc., *86*, 2134 (1964)
25. Brown, T. L., Ladd, J. A., and Newman, G. N.: J. Organomet. Chem., *3*, 1 (1965)
26. Lewis, H. L., and Brown, T. L.: J. Am. Chem. Soc., *92*, 4664 (1970)
27. Margerison, D., and Newport, J. P.: Trans. Faraday Soc., *59*, 2058 (1963)
28. Wittig, G., Meyer, F. J., and Lange, G.: Ann., *571*, 167 (1951)
29. Margerison, D., and Pont, J. D.: Trans. Faraday Soc., *67*, 353 (1971)
30. Bywater, S., and Worsfold, D. J.: J. Organomet. Chem., *10*, 1 (1967)
31. Weiner, M., Vogel, C., West, R.: Inorg. Chem., *1*, 654 (1962)
32. Baney, R. H., and Krager, R. J.: Inorg. Chem. *3*, 1967 (1964)
33. Hartwell, G. E., and Brown, T. L.: Inorg. Chem. *3*, 1656 (1964)
34. Fraenkel, G., Beckenbaugh, W. E., and Yang, P. P.: J. Am. Chem. Soc., *98*, 6878 (1976)
35. Glaze, W. H., and Freeman, C. H.: J. Am. Chem. Soc., *91*, 7198 (1969).
36. Glaze, W. H., Hanicak, J. E., Moore, M. L., and Chaudhuri, J.: J. Organomet. Chem., *44*, 39 (1972)
37. Exner, M. M., Waack, R., and Steiner, E. C.: J. Am. Chem. Soc., *95*, 7009 (1973)
38. Hsieh, H. L.: J. Polym. Sci., *A-3*, 153, 163 (1965)
39. Selman, C. M., and Hsieh, H. L.: Polym. Lett., *9*, 219 (1971)
40. Hsieh, H. L., and Glaze, W. H.: Rubber Revs., *43*, 22 (1970)
41. Makowski, H. L., and Lynn, M.: J. Macromol. Chem; *1*, 443 (1966)
42. Morton, M., Fetters, L. J., Pett, R. A., and Meier, J. F.: Macromolecules *3*, 327 (1970)
43. Worsfold, D. J., and Bywater, S.: Macromolecules, *5*, 393 (1972)
44. Fetters, L. J., and Morton, M.: Macromolecules *7*, 552 (1974)
45. Morton, M., Bostick, E. E., and Livigni, R. A.: Rubber Plast. Age, *42*, 397 (1961)
46. Morton, M., Fetters, L. J., and Bostick, E. E.: J. Polym. Sci., Part C, *1*, 311 (1961)
47. Morton, M., and Fetters, L. J.: J. Polym. Sci., Part A, *2*, 3311 (1964)
48. Al-Jarrah, M. M., and Young, R. N.: Polymer *21*, 119 (1980)

49. Worsfold, D. J., and Bywater, S.: Can. J. Chem., *42*, 2884 (1964)
50. Hernandez, A., Semel, J., Broecker, H. C., Zachmann, H. G., and Sinn, H.: Makromol. Chem., Rapid Commun., *1*, 75 (1980)
51. Johnson, A. F., and Worsfold, D. J.: J. Polym. Sci., Part *A-3*, 449 (1965)
52. Fetters, L. J., and Young, R. N.: in *Anionic Polymerization: Kinetics, Mechanisms, and Synthesis*, McGrath, J. E., ed., ACS Symp. Series 166, Am. Chem. Soc., Washington, D. C., 1981, p. 95
53. Panek, E. J., and Whitesides, G. M.: J. Am. Chem. Soc., *94*, 8768 (1972)
54. West, P., Purmort, J. I., and McKinley, S. V.: J. Am. Chem. Soc., *90*, 797 (1968)
55. Brubaker, G. R., and Beak, P.: J. Organomet. Chem., *136*, 147 (1977)
56. West, P., and Waack, R.: J. Am. Chem. Soc., *89*, 4395 (1967)
57. Das, R., and Wilkie, C. A.: J. Am. Chem. Soc., *94*, 4555 (1972)
58. Halaska, V., and Lochmann, L.: Coll. Czech. Chem. Commun., *38*, 1780 (1973)
59. Viswanathan, C. T., and Wilkie, C. A.: J. Organomet. Chem., *54*, 1 (1973)
60. Rodinov, A. N., Shigorin, D. N., Talaleva, T. L., Tsareva, G. V., and Kocheskov, K. A.: Zhur. fiz. Khim., *40*, 2265 (1966). Chem. Abst., *66*, 28268e (1967)
61. Das, K. S., Feld, M., and Szwarc, M.: J. Am. Chem. Soc., *82*, 1506 (1960)
62. Fetters, L. J.: J. Polym. Sci., B, *2*, 425 (1964)
63. Hartwell, G. E., and Brown, T. L.: Inorg. Chem., *5*, 1257 (1966)
64. Halaska, V., Lochmann, L., and Lim, P.: Coll. Czech. Chem. Commun., *33*, 3245 (1968)
65. Streitwieser, A., and Padgett, W. M.: J. Phys. Chem., *68*, 2919 (1964)
66. Kimura, B. Y., and Brown, T. L.: J. Organomet. Chem., *26*, 57 (1971)
67. Clark, T., Schleyer, P. V., and Pople, J. A.: J. Chem. Soc., Chem. Commun., 137 (1978)
68. Baird, N. C., Barr, R. F., and Datta, R. K.: J. Organomet. Chem., *59*, 65 (1973)
69. Guest, M. F., Hillier, I. H., and Sanders, V. R.: J. Organomet. Chem., *44*, 59 (1972)
70. Brown, T. L.: J. Organomet. Chem., *5*, 191 (1966)
71. Seitz, L. M., and Brown, T. L.: J. Am. Chem. Soc., *88*, 2174 (1966)
72. Darensburg, M. Y., Kimura, B. Y., Hartwell, G. D., and Brown, T. L.: J. Am. Chem. Soc., *92*, 1237 (1970)
73. Berkowitz, J., Bafus, D. A., and Brown, T. L.: J. Phys. Chem., *65*, 1380 (1961)
74. Erussalimsky, G. B., and Kormer, V. A., Eur. Polym. J., *16*, 463 (1980)
75. Erussalimsky, G. B., and Kormer, V. A.: Vysokomol. Soedin, *B19*, 169 (1977)
76. Roovers, J. E. L., and Bywater, S.: Polymer *14*, 594 (1973)
76a. Szwarc, M.: Adv. Polym. Sci., *47*, 1 (1982)
77 Szwarc, M.: J. Polym. Sci., *18*, 493 (1980)
78. Szwarc, M.: in *Anionic Polymerization: Kinetics, Mechanism, and Synthesis*, McGrath, J. E., ed., ACS Symp. Series 166, Am. Chem. Soc., Washington, D.C., 1981, p. 1
79. Meier, J. F.: Ph. D. thesis, The University of Akron, 1963 (available from University Microfilms, Ann Arbor, Mich.)
80. Quirk, R. P., and Kester, D. E., and Delaney, R. D.: J. Organomet. Chem., *59*, 45 (1973)
81. Quirk, R. P., and Kester, D. E.: J. Organomet. Chem., *C 23*, 72 (1974)
82. Quirk, R. P., and Kester, D. E.: J. Organomet. Chem., *127*, 111 (1977)
83. Quirk, R. P., and McFay, D.: Makromol. Chem., Rapid Commun., *1*, 71 (1980)
84. Arnett, E. M., Bentrude, W. G., Burke, J. J., and Duggleby, P. McC.: J. Am. Chem. Soc., *87*, 1541 (1965).
85. Brown, T. L., Dickerhoff, D. W., Bafus, D. A., and Morgan, G. L.: Rev. Sci. Instr., *33*, 491 (1962).
86. Gilman, H., and Cartledge, F. K.: J. Organomet. Chem., *2*, 447 (1964)
87. *Polyamine-Chelated Alkali Metal Compounds*, Langer, A. W., ed., Advances in Chemistry Series, *130*, Am. Chem. Soc., Washington, D.C., 1974
88. Quirk, R. P.: in *Anionic Polymerization: Kinetics, Mechanisms, and Synthesis*, McGrath, J. E., ed., ACS Symposium Series 166, Am. Chem. Soc., Washington, D.C., 1981, p. 117
89. Quirk, R. P., and McFay, D.: J. Polym. Sci., Polym. Chem. Ed., *19*, 1445 (1981)
90. Dumas, S., Marti, V., Sledz, J., and Schue, F.: J. Polym. Sci., Polym. Lett. Ed., *16*, 81 (1978)
91. Antkowiak, T. A., Oberster, A. E., Halasa, A. F., and Tate, D. P.: J. Polym. Sci., A-1, *10*, 1319 (1972)

 92. Helary, G., and Fontanille, M.: Eur. Polym. J., *14*, 345 (1978)
 93. Milner, R., and Young, R. N.: unpublished observations.
 94. Worsfold, D. J., and Bywater, S.: Can. J. Chem. *38*, 1891 (1960)
 95. Burnett, G. M., and Young, R. N.: Eur. Polym. J., *2*, 329 (1966)
 96. Stupfer, R., Kaempf, B., and Tanielian, C.: J. Polym. Sci., Polym. Chem. Ed., *16*, 453 (1978)
 97. Evans, A. G., and George, D. B.: J. Chem. Soc., 4653 (1961)
 98. Casling, R. A. H., Evans, A. G., and Rees, N. H.: J. Chem. Soc., B, 519 (1966)
 99. Mechin,R., Kaempf, B., and Tanielian, C.: Eur. Polym. J., *13*, 493 (1977)
100. Roovers, J. E. L., and Bywater,S.:Macromolecules*1*,328(1968)
101. Guyot, A., and Vialle, J.: J. Polym. Sci., Part B *6*, 403 (1968)
102. Guyot, A., and Vialle, J.: J. Macromol. Sci., Chem., *A4*, 79 (1970)
103. Schué, F., and Bywater, S.: Macromolecules *2*, 458 (1969)
104. Morton, M., Pett, R. A., and Fetters, L. J.: Macromolecules *3*, 333 (1970)
105. Smirnowa, N., Sgonnik, V., Kalninsch, K., and Erussalimsky, B.: Makromol. Chem. *178*, 733 (1977)
106. Hsieh, H.: J. Polym. Sci., *A3*, 163 (1965)
107. Roovers, J. E. L., and Bywater, S.: Macromolecules, *8*, 251 (1975)
108. Fetters, L. J., and Ruppert, J.: unpublished results
109. Burley, J. A., and Young, R. N.: Chem. Comm., 991 (1970)
110. O'Driscoll, K. F., Ricchezza, E. N., and Clark, J. E.: J. Polym. Sci., Part A, *3*, 3241 (1965)
111. Smart, J. B., Emerson, M. T., and Oliver, J. P.: J. Am. Chem. Soc., *88*, 4101 (1966)
112. Smart, J. B., Hogan, R., Scherr, P. A., Emerson, M. T., and Oliver, J. P.: J. Organomet. Chem., *64*, 1 (1974)
113. Hall, J. L.: Ph. D. Thesis, The University of Akron, 1962, (available from University Microfilms, Ann Arbor, Mich.)
114. Tanlak, T., Ahmad, A., Treybig, M. N., and Anthony, R. G.: J. Appl. Polym. Sci., *22*, 315 (1978)
115. Ceausescu, E., Bordeianu, R., Cerchez, I., Busdugan, E., Ilie, V., and Ghioca, P.: Acta Polym., *34*, in press (1983)
116. Bartlett, P. D., Goebel, C. V., and Weber, W. P.: J. Am. Chem. Soc., *91*, 7425 (1969)
117. Glaze, W. H., Hanicak, J. E., Berry, D. J., and Duncan, D. P.: J. Organomet. Chem., *44*, 49 (1972)
118. Glaze, W. H., Lin, J., and Felton, E. G.: J. Org. Chem., *30*, 1258 (1965)
119. Cubbon, R. C. P., and Magerison, D.: Proc. Roy. Soc., *268A*, 260 (1962)
120. Spirin, Y. L., Gantmakker, A. R., and Medvedev, S. S.: Dokl. Akad. Nauk., SSSR, *146*, 368 (1962); Chem. Abst. *58*, 1537d (1963)
121. Spirin, Y. L., Polykov, D. K., Grantmakker, A. R., and Medvedev, S. S.: J. Polym. Sci., *53*, 233 (1961)
122. Morton, M., Pett, R. A., and Fellers, J. F.: Preprints IUPAC Macromol. Symp., Tokyo *1*, 69 (1966)
123. Alvarino, J. M., Bello, A., and Guzman, G. M.: Eur. Polym. J., *8*, 53 (1972)
124. Sinn, H., Lundborg, C., Onsager, O. T.: Makromol. Chem., *70*, 222 (1964)
125. Morton, M., Bostick, E. E., and Fetters, L.J.:J.Polym. Sci., Part C *1*, 311 (1963)
126. Morton, M., Bostick, E. E., Livigni, R. A., and Fetters, L. J.: J. Polym. Sci., Part A, *1*, 1735 (1963)
127. Fetters, L. J.: J. Res. Nat. Bur. Stds., Sect. A *69*, 159 (1965)
128. Kuntz, I.: J. Polym. Sci., Part A *2*, 2827 (1964)
129. Francois, B., Sinn, V., and Parrod, J.: J. Polym. Sci. Part C, *4*, 375 (1964)
130. Crammond, D. N., Lawry, P. S., and Urwin, J. R.: Eur. Polym. J., Eur. Polym. J., *2*, 107 (1966)
131. Shamanim, V. V., Melenevskaya, E. Yu., and Zgonnik, V. N.: Acta Polym., *33*, 175 (1982)
132. Hadjichristidis, N., and Roovers, J. E. L.: Polym. Sci., Polym. Phys. Ed., *12*, 2521 (1974)
133. Worsfold, D. J.: J. Polym. Sci., Polym. Phys. Ed., *20*, 99 (1982)
134. Schué, F., Worsfold, D. J., and Bywater, S.: J. Polym. Sci., Part B, *7*, 821 (1969); Bull. Soc. Chim., Fr., 271 (1970); Macromolecules *3*, 509 (1970)
135. Morton, M., Sanderson, R. D., and Sakata, R.: Polym. Letts., *9*, 61 (1971)
136. Glaze, W. H., and Jones, P. C.: J. Chem. Soc.: D, 1434 (1969)
137. Naylor, F. E., Hsieh, H. L., and Randall, J. C.: Macromolecules *3*, 486 (1970)

138. Schué, F., Worsfold, D. J., and Bywater, S.: Macromolecules 3, 509 (1970)
139. Bywater, S., Worsfold, D. J., and Hollingsworth, G.: Macromolecules 5, 389 (1972)
140. Halasa, A. F.,Schulz,D. N., Tate, D. P., and Mochel, V. D.: Adv. Organomet. Chem., 18,55
     (1980)
141. Morton, M., Sanderson, R. D., and Sakata, R.: Macromolecules 5, 389 (1972)
142. Morton, M., Sanderson, R. D., Sakata, R., and Falvo, L. A.: Macromolecules 6, 186 (1973)
143. Halasa, A. F., Mochel, V. D., and Fraenkel, G.: in Anionic Polymerization: Kinetics: Mechanism,
     and Synthesis, McGrath, J. E., ed., ACS Symp. Series 166, Am. Chem. Soc., Washington, D.C.,
     1981, p. 367
144. Fetters, L. J.: unpublished results
145. French, D. M.: Rubber Chem. Tech., 42, 71 (1969)
146. Morton, M., Fetters, L. J., Inomata, J., Rubio, D. C., and Young, R. N.: Rubber and Chem.
     Tech., 49, 303 (1976)
147. Campos-Lopez, E., Leon-Gross, A., and Ponce-Velez, M. A.: J. Polym. Sci., Polym. Chem. Ed.,
     11, 3021 (1973)
148. Utracki, L. A., and Roovers, J. E. L.: Macromolecules 6, 366 (1973)
149. Graessley, W. W., Masuda, R., Roovers, J. E. L., and Hadjichristidis, N.: Macromolecules 9,
     127 (1976)
150. Doi, M., and Kuzuu, N. Y.: J. Polym. Lett. Ed., 18, 775 (1980)
151. Hsieh, H. L., and Kitchen, A. G.: in: Initiation of Polymerization, Bailey, Jr., F. E., ed., ACS
     Symp. Series 212, Am. Chem. Soc., Washington, D.C., 1983, p. 291
152. Milner, R., Young, R. N., and Luxton, A. R.: Polymer 24, 543 (1983)
153. O'Driscoll, K., and Patsiga, R.: J. Polym. Sci., Part A, 3, 1037 (1965)
154. Makowski, H. S., Lynn, M., and Bogard, A. N.: J. Macromol. Sci., Chem., A2, 665 (1968)
155. Kaspar, M., and Trekoval, J.: Coll. Czech, Chem. Commun., 41, 1356 (1976); 45, 1047
     (1980)
156. Bywater, S., and Worsfold, D. J.: Can. J. Chem., 40, 1565 (1962)
157. Geerts, J., Van Beylen, M., and Smets, G.: J. Polym. Sci., A-1, 7, 2805 (1969)
158. Yamagishi, A., Szwarc, M., Tung, L., and Lo, G.-Y. S.: Macromolecules 11, 607 (1978)
159. Bernard, D. A., and Noolandi, J.: Macromolecules 15, 1553 (1982)
160. Wang, H. C., and Szwarc, M.: Macromolecules 13, 452 (1980)
161. Szwarc, M., and Wang, H. C.: Macromolecules, 15, 208 (1982)
161a. Kminek, I., Kaspar, M., and Trekoval, J.: Czech. Chem. Commun., 46, 2371 (1981)
162. Fetters, L. J., and Young, R. N.: Macromolecules, 15, 206 (1982)
163. Szwarc, M.: Carbanions, Living Polymers, and Electron Transfer Processes, Interscience
     Publishers, New York, 1968, p. 510
164. Geerts, G., Van Beylen, M., and Smets, G.: J. Polym. Sci., A-1, 7, 2859 (1969)
165. Kminek, I., Kaspar, M., and Trekoval, J.: Coll. Czech. Chem. Commun., 46, 1124 (1981)
166. Davidjan, A., Nikolaew, N., Sgonnik, V., Krasikow, V., Belenkii, B., and Erussalimsky, B.:
     Makromol., Chem., 182, 917 (1981)
167. Ziegler, K., and Gellert, H. C.: Ann. 567, 195 (1950)
168. Langer, A. W.: Trans. N.Y.Acad.Sci. 27, 741(1965)
169. Eberhardt,G. G., and Butte, W. A.: J. Org. Chem. 29, 2928 (1964)
170. Eberhardt, G. G., and Davis, W. R.: J. Polym. Sci. A3, 3753 (1965)
171. Hay, J. N., McGabe, J. R., and Robb, J. C.: JCS Faraday Trans 1, 68, 1 (1972)
172. Hay, J. N., Harris, D. S., and Wiles, M.: Polymer 17, 613 (1976)
173. Magnin, H., Rodriquez, F., Abadie, M., and Schué, F. J.: J. Polym. Sci., Polym. Chem. Ed.,
     15, 857 (1977)
174. Rodriquez, F., Abadie, M., and Schué, F.: Europ. Polym. J., 12, 17 (1976)
175. Aldissi, M., Schué, F., Geckeler, K., and Abadie, M.: Makromol. Chem., 181, 1413 (1980)
176. Bartlett, P. D., Friedmann, S., and Stiles, M.: J. Amer. Chem. Soc., 75, 1771 (1953)
177. Crassous, G., Abadie, M., and Schué, F.: Europ. Polym. J., 15, 767 (1979)
178. Helary, G., and Fontanille, M.: Polymer Bull., 3, 159 (1980)
179. Worsfold, D. J., and Bywater, S.: J. Phys. Chem., 70, 162 (1966)
180. Hay, J. N., and McCabe, J. F.: J. Polym. Sci., Polym. Chem. Ed., 10, 3451 (1972)
181. Milner, R., and Young, R. N.: Polymer 23, 1636 (1982)
182. Vinogradova, L.V., Nikolayev, N. I., Sgonnik, V. N.: Vysokomol. Soedin, A18, 1756 (1976)

183. McElroy, B. J., and Merkley, J. H.: (to Lithium Corp.) U.S. Patent, 3,678,1217 (1972)
184. Quack, G., and Fetters, L. J.: Macromolecules, *11*, 369 (1978)
185. Luxton, A. R., Burrage, M., Quack, G., and Fetters, L. J.: Polymer *22*, 382 (1981)
186. Smirnowa, N., Sgonnik, V., Kalninsch, K., and Erussalimsky, B.: Makromol. Chem., *178*, 773 (1977)
187. Davidjan, A., Nikolaew, N., Sgonnik, N., Belenkii, B., Nesterow, V., and Erussalimsky, B.: Makromol. Chem., *177*, 2469 (1976)
188. Collet-Marti, V., Dumas, S., Sledz, J., and Schué, F.: Macromolecules, *15*, 251 (1982)
189. Davidjan, A., Nikolaew, N., Sgonnik, N., Belenkii, B., Nesterow, V., Krasikow, V., and Erussalimsky, B.: Makromol. Chem., *179*, 2155 (1978)
190. Cheminat, A., Friedmann, G., and Brini, M.: J. Polym. Sci., Polym. Chem. Ed., *17*, 2865 (1979)
191. Dyball, C. J., Worsfold, D. J., and Bywater, S.: Macromolecules, *12*, 819 (1979)
192. Garton, A., and Bywater, S.: Macromolecules *8*, 694 (1975)
193. Garton, A., Chaplin, R. P., and Bywater, S.: Europ. Polym. J., *12*, 697 (1976)
194. Bywater, S., Johnson, A. F., and Worsfold, D. J.: Can. J. Chem., *42*, 1255 (1964)
195. Glaze, W. G., Hanicak, J. E., Chandhuri, J., Moore, M. L., and Duncan, D. P.: J. Organometal. Chem., *51*, 13 (1973)
196. Bywater, S., and Worsfold, D. J.: J. Organometal. Chem., *159*, 229 (1978)
197. Brownstein, S., Bywater, S., and Worsfold, D. J.: Macromolecules, *6*, 715 (1973)
198. Lachance, P., and Worsfold, D. J.: J. Polym. Sci., Polym. Chem. Ed., *11*, 2295 (1973)
199. Morton,M., Falvo, L. A., and Fetters, L. J.: J. Polym. Sci., Polymer Letters *10*, 561 (1972)
200. Morton, M., and Falvo, L. A.: Macromolecules *6*, 190 (1973)
201. Bywater, S.: Polymer Journal, *12*, 569 (1980)
202. Takahashi, K., Takaki, M., and Asami, R.: Org. Magn. Reson., *3*, 539 (1971)
203. Bywater, S., and Worsfold, D. J.: J. Organometal. Chem., *33*, 273 (1971)
204. Takahashi, K., Takaki, M., and Asami, R.: J. Phys. Chem., *75*, 1062 (1971)
205. Sandel, V. R., and Freedman, H. H.: J. Amer. Chem. Soc., *85*, 2328 (1962)
206. Takahashi, K., Yamada, K., Wakata, K., and Sami, R.: Org. Magn. Reson., *6*, 62 (1974)
207. Yudin, V. P.: Vysokomol. Soed. *A20*, 1001 (1978) translated in Polymer Science USSR *20*, 1126 (1979)
208. Gerbert, W., Hinz, J., and Sinn, H.: Makromol. Chem., *144*, 97 (1971)
209. Morton, M., and Rupert, J.: in *Initiation of Polymerization*, Bailey, Jr., F. E., Ed., ACS Symposium Series No. 212, American Chemical Soc., Washington, D.C., 1983, p. 283
210. Suzuki, T., Tsuji, Y., and Takegami, Y.: Macromolecules *11*, 639 (1978)
211. Inomata, J., Makromol. Chem., *135*, 113 (1970)
212. Worsfold, D. J., and Bywater, S.: Macromolecules *11*, 582 (1978)
213. Jenner, G., and Khalilpour, A.: Europ. Polùm. J., *12*, 105 (1976)
214. Beilin, S. I., Dolgoplask, B. A., and Tinyakova, E. I.: Europ. Polym. J., *11*, 409 (1975)
215. Glaze, W. H., Reinarz, R. B., and Moore, M. L.: J. Polymer Sci., Polymer Letters *15*, 141 (1977)
216. Aubert, P., Sledz, J., Schué, F., and Prd'homme, J.: Europ. Polym. J., *16*, 361 (1980)
217. Uraneck, C. A.: J. Polym. Sci., *A1 9*, 2273 (1971)
218. Salle, R., and Pham, Q. T.: J. Polym. Sci., *A1 15*, 1799 (1977)
219. Salle, R., Essel, A., Gole, J., and Pham, Q. T.: Polym. Sci., *A1*, *13*, (1975)
220. Yuki, H., Okamoto, Y., and Takano, H.: Polymer J., *2*, 663 (1971)
221. Roy, N., and Prud'homme, J.: Macromolecules *8*, 78 (1975)
222. Glague, A. D. H., van Broekhoven, J. A. M., and Blaauw, L. P.: Macromolecules *7*, 348 (1974)
223. Elgert, K. F., Quack, G., and Stutzel, B.: Makromol. Chem. *175*, 1955 (1974)
224. Mochel, V. D.: J. Polym. Sci., *A1 10*, 1009 (1972)
225. Tanaka, Y., Sato, H.: J. Polym. Sci., Polym. Letters Ed., *16*, 472 (1978)
226. Beebe, D. H.: Polymer *19*, 231 (1978)
227. Medvedev, M. Y.: Vysokomol. Soedin, *B 11*, 395 (1967)
228. Bywater, S., and Worsfold, D. J.: Can. J. Chem., *45*, 1821 (1967)
229. Kow, C., Hadjichristidis, N., Morton, M., and Fetters, L. J.: Rubber Chem. Technol. *55*, 245 (1982)

230. Essel, A., Salle, R., Gole, J., and Pham, Q. T.: J. Polym. Sci., Polym. Chem. Ed., *13*, 1847 (1975)
231. Essel, A., and Pham, Q. T.: J. Polym. Sci., *A1*, *10*, 2793 (1972)
232. Mays, J., and Fetters, L. J.: unpublished results
233. Bywater, S., Patmore, D. J., and Worsfold, D. J.: J. Organomet. Chem., *135*, 145 (1977)
234. Matsuzaki, K., Uryu, T., Seki, T., Osada, K., and Kawamura, T.: Makromol. Chem., *176*, 3051 (1975)
235. Randall, J. C.: J. Polym. Sci., Polym. Chem. Ed., *13*, 889 (1975)
236. Uryu, T., Kawamura, T., and Matsuzaki, K.: J. Polym. Sci., Polym. Chem. Ed., *17*, 2019 (1979)
237. Suparno, S., Lacoste, J., Raynal, S., Sledz, J., and Schué, F.: Polymer J., *13*, 313 (1981)
238. Wicke, R., and Elgert, K.-F.: Makromol. Chem., *178*, 3075 (1977)
239. Wicke, R., and Elgert, K.-F.: Makromol. Chem., *178*, 3085 (1977)
240. Hostalka, H., Figini, R. V., and Schulz, G. V.: Makromol. Chem., *71*, 198 (1964)
241. Hostalka, H., and Schulz, G. V.: J. Polym. Sci., *B3*, 175 (1965)
242. Hostalka, H., and Schulz, G. V.: Z. physik. Chem., *45*, 286 (1965)
243. Bhattacharyya, D. N., Lee, C. L., Smid, J., and Szwarc, M.: Polymer *5*, 54 (1964)
244. Bhattacharyya, D. N., Lee, C. L., Smid, J., and Szwarc, M.: J. Phys. Chem. *69*, 612 (1965)
245. Arest-Yakubovich, A. A., and Medvedev, S. S.: Dokl. Akad. Nauk SSSR, *159*, 1066 (1964)
246. Bandermann, F., and Sinn, H.: Makromolek. Chem., *96*, 150 (1966)
247. Vinogradova, L. V., Zgonnik, V. N., Nikolaev, N. I., and Tsvetanov, Kh. B.: Eur. Polym. J., *15*, 545 (1979)
248. Vinogradova, L. V., Zgonnik, V. N., Nikolaev, N. I., and Vetchinova, E. P.: Eur. Polym. J., *16*, 799 (1980)
249. Garton, A., and Bywater, S.: Macromolecules *8*, 697 (1975)
250. Kakova, G. V., and Korotkov, A. A.: Dokl. Akad. Nauk SSSR, *119*, 982 (1958); Rubber Chem. Tech., *33*, 623 (1960)
251. Wang, I.-C., Mohajer, Y., Ward, T. C., Wilkes, G. L., and McGrath, J. E.: in *Anionic Polymerization: Kinetics, Mechanisms, and Synthesis*, McGrath, J. E., ed., ACS Symp. Series 166, Am. Chem. Soc., Washington, D.C., 1981, p. 529
252. Furukawa, J., Saegusa, T., and Irako, K.: Kogaku Zasshi, *65*, 2029 (1962); Chem. Abstr., *58*, 11553 (1963)
253. Morton, M., and Ells, F. R.: J. Polym. Sci., *61*, 25 (1962)
254. Korotkov, A. A.: Angew. Chem., *70*, 85 (1958)
255. Korotkov, A. A., and Chesnokova, N. N.: Polym. Sci. USSR, *2*, 284 (1961)
256. Spirin, Yu. L., Polyakov, D. K., Gantmakher, A. R., and Medvedev, S. S.: Dokl. Akad. Nauk SSR, *139*, 899 (1961); Proc. Acad. Sci. (USSR), *139*, 778 (1961)
257. Johnson, A. F., and Worsfold, D. J.: Makromol. Chem., *85*, 273 (1965)
258. Mochel, V. D.: Rubber Chem. Tech., *40*, 1200 (1967)
259. Kuntz, I.: Polym. Sci., *54*, 569 (1961)
260. Aggarwall, S. L., Livigni, R. A., Marker, L. F., and Dudek, T. J.: in *Block and Graft Copolymers*, Burke, J. J., and Weiss, V., ed., Syracuse University Press, Syracuse, N.Y., 1973, p. 157
261. Rakova, G. V., and Korotkov, A. A.: Polym. Sci. USSR, *3*, 990 (1962)
262. Spirih, Yu. L., Polyakov, D. K., Gantmakher, A. R., and Medvedev, S. S.: Polym. Sci. USSR, *3*, 233 (1963)
263. Worsfold, D. J.: J. Polym. Sci. A-1, *5*, 2783 (1967)
264. Sivola, A.: Acta Polytechnica Scandinavia, *134*, 1 (1977)
265. Chen, J., and Fetters, L. J.: Poly. Bull., *4*, 275 (1981)
266. O'Driscoll, K. F., and Patsiga, R.: J. Polym. Sci., Part A, *3*, 1037 (1965)
267. Ref. 163, p. 540
268. O'Driscoll, K. F., and Kuntz, I.: J. Polym. Sci., *61*, 19 (1962)
269. Young, R. N., and Fetters, L. J.: Macromolecules, *11*, 899 (1978)
270. Quack, G., Fetters, L. J., Hadjichristidis, N., and Young, R. N.: Ind. Eng. Chem.: Prod. Res. Dev., *19*, 587 (1980)
271. Yamagishi, A., and Szwarc, M.: Macromolecules *11*, 504 (1978)
272. Busson, R., and van Beylen, M.: Macromolecules, *10*, 1320 (1977)

273. Kelley, D. J., and Tobolsky, A. V.: J. Am. Chem. Soc., *81*, 1597 (1981)
274. Robertson, R. E., and Marion, L.: Can. J. Res., *26B*, 657 (1948)
275. Bower, F. M., and McCormick, H. W.: J. Polym. Sci., Part A, *1*, 1749 (1963)
276. Brooks, B. W.: Chem. Commun., *68* (1967)
277. Higginson, W. C. E., and Wooding, N. S.: J. Chem. Soc., 760 (1952)
278. Gatzke, A.: J. Polym. Sci. Part A-1, 7, 2281 (1969)
279. Kume, S., Takahashi, G., Nishikawa, G., Hatano, M., and Kambara: Makromol. Chem., *84*, 137, 147 (1965); ibid., *98*, 109 (1966)
280. Kume, S.: Makromol. Chem., *98*, 120 (1966)
281. Luxton, A. R.: Rubber Revs., *54*, 596 (1981)
282. Ziegler, K., and Gellert, H. G.: Ann. Chem., *567*, 179 (1950)
283. Bach, R. O., Ellestad, R. B., Kamienski, C. W., Wasson, J. R.: in *Encyclopedia of Chemical Technology*, Vol. 14, 3rd Ed., John Wiley and Sons, Inc., New York, 1981, p. 448
284. Bryce-Smith, D.: J. Chem. Soc., 1712 (1955)
285. Finnegan, R. A., and Kutta, H. W.: J. Org. Chem., *30*, 4138 (1965)
286. Antkowiak, T. A.: Polym. Preps, Amer. Chem. Soc., Div. Polym. Chem., *12*, (2), 393 (1971)
287. Nentwig, W., and Sinn, H.: Makromol. Chem., Rapid Commun. *1*, 59 (1980)
288. Medina, S., Fetters, L. J., and Young, R. N.: unpublished observations
289. Margerison, D., and Nyss, V. A.: J. Chem. Soc., (C), 3065 (1968)
290. Benkeser, R. A., Hooz, J., Liston, T. V., and Trevillyan, A. E.: J. Am. Chem. Soc., *85*, 3985 (1963)
291. Spach, G., Levy, M., and Szwarc, M.: J. Chem. Soc., 355 (1962)
292. Levy, M., Szwarc, M., Bywater, S., and Worsfold, D. J.: Polymer *1*, 515 (1960)
293. Schmitt, B. J., and Schulz, G. V.: Makromol. Chem., *121*, 184 (1969)
294. Komiyama, J., Böhm, L. L., and Schulz, G. V.: Makromol. Chem., *148*, 297 (1971)
295. Böhm, L. L., Chmelir, M., Löhr, G., Schmitt, B. J., and Schulz, G. V.: Adv. Polym. Sci., *9*, 1 (1972)
296. Schmitt, B. J.: Makromol. Chem., *156*, 243 (1972)
297. Podolsky, A. F., and Taran, A. A.: J. Polym. Chem., Polym. Chem. Ed., *12*, 2187 (1974)
298. Comyn, J., and Glasse, M. D.: J. Polym. Sci., Polym. Lett. Ed., *18*, 703 (1980)
299. Decker, D., Roth, J.-P., and Franta, E.: Makromol. Chem., *162*, 279 (1972)
300. Okamoto, A., Watanake, E., and Mita, I.: J. Polym. Sci., Polym. Chem. Ed., *17*, 2483 (1979)
301. Comyn, J., and Glasse, M. D.: Makromol. Chem., *175*, 695 (1974)
302. Williams, R. L., and Richards, D. H.: Chem. Comm. 414 (1967)
303. Richards, D. H., and Scilly, N. F.: Chem. Comm. 1641 (1968)
304. Richards, D. H., and Williams, R. L.: J. Polym. Sci., Polym. Chem. Ed., *11*, 89 (1973)
305. Lee, C. L., Smid, J., and Szwarc, M.: J. Phys. Chem., *66*, 904 (1962)
306. Vrancken, A., Smid, J., and Szwarc, M.: Trans. Faraday Soc., *58*, 2036 (1962)
307. Szwarc, M.: Adv. Polym. Sci., *4*, 457 (1967)
308. Ades, D., Fontanille, M., and Leonard, J.: Can. J. Chem. *60*, 564 (1982)
309. Comyn, J., and Glasse, M. D.: J. Polym. Sci., Polym. Chem. Ed., *21*, 227 (1983)
310. Comyn, J., and Glasse, M. D.: J. Polym. Sci., Polym. Chem. Ed., *21*, 209 (1983)
311. Burley, J. W., and Young, R. N.: J. Chem. Soc. C, 3780 (1971)
312. Audureau, J., Fontanille, M., and Sigwalt, P.: C. R. Acad. Sci. Ser. C., *275*, 1487 (1972)
313. Vinogradova, L. V., Nikolaev, N. I., Sgonnik, V. N., Erussalimsky, B. L., Sinitsina, G. V., Tsuetanov, Ch. B., and Panayotov, I. M.: Eur. Polym. J., *17*, 517 (1981)
314. Morton, M., and Fetters, L. J.: Macromol. Rev., *2*, 71 (1967)
315. Fetters, L. J.: J. Polym. Sci., Part C, *26*, 1 (1969)
316. Bywater, S.: Prog. Polym. Sci., *4*, 27 (1974)
317. Ref. 163, p. 96
318. Wyman, D. P., Allen, V. R., and Altares, T.: J. Polym. Sci. *A*, *2*, 4545 (1964)
319. Mansson, P.: J. Polym. Sci., Polym. Chem. Ed., *18*, 1945 (1980)
320. Quirk, R. P., and Chen, W.-C.: Makromol. Chem., *183*, 2071 (1982)
321. Jorgenson, M. J.: Org. React., *18*, 1 (1970)
322. *Polymer Handbook*, 2nd Ed., J. Brandrup and E. H. Immergut, Ed., Wiley-Interscience, New York, 1975, p. IV–167

323. Brody, H., Richards, D. H., and Szwarc, M.: Chem. Ind. (London), 1473 (1958)
324. Komatsu, K., Nishioka, A., Ohshima, N., Takahashi, M., and Hara, H.: U.S. Patent 4,083,834 (April 11, 1978)
325. Pannell, J.: Polymer, *12*, 547 (1971)
326. Burgess, F. J., and Richards, D. H.: Polymer, *17*, 1020 (1976)
327. Burgess, F. J., Cunliffe, A. V., Richards, D. H., and Sherrington, D. C.: Polym. Lett., *14*, 471 (197
328. Russell, G. A., and Lamson, D. W.: J. Am. Chem. Soc., *91*, 3967 (1969)
329. Whitesides, G. M., Panek, E. J., and Stedronsky, E. R.: J. Am. Chem. Soc., *94*, 232 (1972)
330. Quirk, R. P., and McFay, D.: unpublished results
331. Richards, D. H., Salter, D. A., and Williams, R. L.: J. Chem. Soc., Chem. Commun., 38 (1966)
332. Reed, S. F., Jr.: J. Polym. Sci. A-1, *10*, 1187 (1972)
333. Steiner, E. C., Pelletier, R. R., and Trucks, R. O.: J. Am. Chem. Soc., *86*, 4678 (1964)
334. Jerome, R., Teyssie, Ph., and Huynh-Ba, G.: in *Anionic Polymerization: Kinetics, Mechanisms, and Synthesis*, McGrath, J. E., Ed., ACS Symposium Series No. 166, American Chemical Society, Washington, D.C., 1981, p. 212
335. Richards, D. H., and Szwarc, M.: Trans. Faraday Soc., *55*, 1644 (1959)
336. Richards, D. H.: J. Polym. Sci., Polym. Letters, *6*, 417 (1968)
337. Schulz, D. N., Halasa, A. F., and Oberster, A. E.: J. Polym. Sci., Polym. Chem. Ed., *12*, 153 (1974)
338. Schulz, D. N., Sanda, J. C., and Willoughby, B. G.: in *Anionic Polymerization: Kinetics, Mechanisms, and Synthesis*, McGrath, J. E., Ed., ACS Symposium Series No. 166, American Chemical Society, Washington, D.C., 1981, p. 427
339. Morton, M., and Mikesell, S. L.: J. Macromol. Sci.-Chem., *A-7*, 1391 (1973)
340. Schulz, D. N., and Halasa, A. F.: J. Polym. Sci., Polym. Chem. Ed., *15*, 2401 (1977)
341. Hirao, A., Hattori, I., Sadagawa, T., Yamaguchi, K., and Nakahama, S.: Makromol. Chem., Rapid Commun., *3*, 59 (1982)
342. Koenig, R., Riess, G., and Banderet, A.: Eur. Polym. J., *3*, 723 (1967)
343. Beak, P., and Kokko, B. J.: J. Org. Chem., *47*, 2822 (1982)
344. Quirk, R. P., and Cheng, P. L.: Polym. Prep. *24*, No. 2, 426 (1983)
345. Cheng, T. C., Oberster, A. E., and Halasa, A. F.: J. Polym. Sci., Polym. Chem. Ed., *17*, 1847 (1979)
346. Cheng, T. C.: in *Anionic Polymerization: Kinetics, Mechanisms, and Synthesis*, McGrath, J. E., Ed., ACS Symposium Series No. 166, American Chemical Society, Washington, D.C., 1981,
347. Angood, A. C., Hurley, S. A., and Tait, P. J. T.: J. Polym. Sci., Polym. Chem. Ed., *11*, 2777 (1973); *ibid, 13*, 2437 (1975)
348. Eisenbach, C. D., Schnecko, H., and Kern, W.: Eur. Polym. J., *11*, 699 (1975)
349. Eisenbach, C. D., Schnecko, H., and Kern, W.: Makromol. Chem. Suppl. *1, 151*, 166 (1975)
350. Catala, J. M., Boscato, J. F., Franta, E., and Brossas, J.: in *Anionic Polymerization: Kinetics, Mechanisms, and Synthesis*, McGrath, J. E., Ed., ACS Symposium Series No. 166, American Chemical Society, Washington, D.C., 1981, p. 483
351. Fetters, L. J., and Firer, E. R.: Polym., *18*, 306 (1977)
352. Quirk, R. P., and Chen, W.-C.: J. Polym. Sci., Polym. Chem. Ed., in press.
353. Russell, G. A.: J. Am. Chem. Soc., *79*, 3871 (1957)
354. Nugent, W. A., Bertini, F., and Kochi, J. K.: J. Am. Chem. Soc., *96*, 4945 (1974)
355. Hagen, A. J., Farrall, M. J., and Fréchet, J. M. J.: Polym. Bull., *5*, 111 (1981)
356. Milkovich, R.: in *Anionic Polymerization: Kinetics, Mechanisms, and Synthesis*, McGrath, J. E., Ed., ACS Symposium Series No. 166, American Chemical Society, Washington, D.C., 1981, p. 41
357. Masson, P., Franta, E., and Rempp, P.: Macromol. Chem., Rapid Commun., *3*, 499 (1982)
358. Chen, J., and Fetters, L. J.: unpublished results
359. Schulz, G. O., and Milkovich, R.: J. Appl. Polym. Sci., *27*, 4773 (1982)
360. De Groof, B., Van Beylen, M., and Szwarc, M.: Macromolecules *8*, 396 (1975)
361. De Smedt, C., and Van Beylen, M.: in *Anionic Polymerization: Kinetics, Mechanisms, and Synthesis*, McGrath, J. E. ed., ACS Symposium Series No. 166, Am. Chem. Soc., Washington, D.C., 1981, p. 127

Prof. J. D. Ferry (Editor)
Received Juni 21, 1983

Anionic Copolymerization of Cyclic Ethers with Cyclic Anhydrides

Jozef Lustoň and František Vašš
Polymer Institute, Slovak Academy of Sciences,
842 36 Bratislava, Czechoslovakia

This article summarizes and analyzes the results obtained for the anionic copolymerization of cyclic ethers with cyclic anhydrides. This reaction is of great practical importance, especially as curing reaction of epoxy resins and is also used for the preparation of linear polyesters with special functional pendant groups.

Our attention was concentrated mainly on the kinetics and mechanisms of copolymerization, the effect of epoxide and anhydride structure in copolymerization, the effect of type and structure of initiatior on the rate, and the course of copolymerization. The probable mechanisms are discussed. The copolymerization in the presence of proton-donor compounds as well as the effect of proton-donors are also considered. For a better understanding of the processes, data and theoretical views on the non-catalyzed reaction are included.

Advances in Polymer Science 56
© Springer-Verlag Berlin Heidelberg 1984

List of Abbreviations

BA	benzoic acid
BuE-PA	monobutylester of phthalic acid
BE-HET	monobenzylester of hexachloroendomethylenetetrahydrophthalic acid
BE-HHPA	monobenzylester of hexahydrophthalic acid
BE-MA	monobenzylester of maleic acid
BE-PA	monobenzylester of phthalic acid
BE-SA	monobenzylester of succinic acid
CA	citraconic anhydride
CHX	cyclohexanol
DMA	N,N-dimethylaniline
DMBA	dimethylbenzylamine
DY 062	high boiling tertiary amine (Ciba Geigy AG)
GA	glutaric anhydride
HEB	2-hydroxy-4-(2,3-epoxypropoxy)benzophenone
HHPA	hexahydrophthalic anhydride
HMTA	hexamethylenetetramine
MA	maleic anhydride
MTHPA	methyltetrahydrophthalic anhydride
NMA	nadic methyl anhydride (methylbicyclo[2.2.1]heptene-2,3-dicarboxylic anhydride isomers)
PA	phthalic anhydride
PGE	phenylglycidyl ether
PLA	palmitic acid
p-CH$_3$ PGE	p-methylphenylglycidyl ether
p-Cl PGE	p-chlorophenylglycidyl ether
p-CO$_2$CH$_3$ PGE	p-acetyloxyphenylglycidyl ether
p-NO$_2$ PGE	p-nitrophenylglycidyl ether
p-OCH$_3$ PGE	p-methoxyphenylglycidyl ether
SA	succinic anhydride
TAP	tris-2,4,6-dimethylaminomethylphenol
TBA	tri-n-butylamine
TEA	triethanolamine
THA	tri-n-hexylamine
THPA	tetrahydrophthalic anhydride
X 157/2378	liquid acid anhydrides of acid number 392 (Ciba Geigy AG)
DGEBA	diglycidyl ether of bisphenol A-based epoxy resin of epoxy content 0.52 eq./100 g
ED-5	Soviet epoxy resin with epoxy content of 20 wt-%

1 Introduction

Polycondensation of diols with dicarboxylic acids or reesterification of dicarboxylic acid esters with diols are the main methods of preparing polyesters. Because of the reversibility of this classical polyester formation, high reaction temperatures, long polycondensation times and low pressures are required to remove low molecular weight reaction products in order to shift the equilibrium to the direction of polyester formation and to obtain sufficiently high molecular weights.

The use of cyclic ethers and cyclic anhydrides in copolyaddition reactions dates back to 1905 when Weinschenk [1] showed a new route for the preparation of polyesters. In a catalyzed reaction, even at low temperatures and short reaction times, a high degree of conversion was achieved. This reaction has been utilized for the curing of epoxy resins. Furthermore, non-catalyzed, anionic, cationic, and coordination — catalyzed copolymerization of cyclic ethers with cyclic anhydrides was also studied. The catalyzed copolyaddition can also be used for the preparation of linear polyesters, synthesis of branched copolymers from poly(vinyl alcohol) [2] or from the copolymer of styrene with acrylic acid [3], modification of pigments [4,5], preparation of special polyesters with light-stabilizing [6,7] or light-sensitive [2,8] groups, preparation of metal-containing polyesters [9,10], and crosslinking of liquid rubbers [11-13].

Several publications dealing with problems of the copolymerization of cyclic ethers with cyclic anhydrides have recently appeared [14-18], but they are not concerned with all aspects of the problems involved.

In this paper, we try to review results obtained from anionic copolymerization of cyclic ethers with cyclic anhydrides. For a better understanding data and theoretical views on non-catalyzed copolymerizations are also included. We concentrate mainly on the kinetics and mechanism of copolymerization and the effect of the type and character of the initiator used. The influence of the epoxide and anhydride structure on copolymerization, of proton donors on the rate and course of copolymerization, and on the molecular weight of the resulting polyesters are also discussed.

2 Non-Catalyzed Reactions of Epoxides with Anhydrides

To understand and comparev the mechanisms and rates of polyester formation in catalyzed copolymerization, processes taking place in the system epoxide-anhydride without any initiator are described. In this review, copolymerization in the absence of compounds that do not occur in the initial reaction mixture is regarded as a non-catalyzed reaction. This means that the presence of alcohols, phenols or acids is not excluded. These compounds may be considered as copolymerization catalysts; however, because of their possible occurrence in the polymerization system they are not regarded as initiators. The presence of OH groups in epoxy compounds, especially in resins where they occur as chlorohydrines (I), monoethers (II), and diethers of glycerol (III)

$$-O-CH_2-CH-CH_2-Cl \qquad -O-CH_2-CH-CH_2 \qquad -O-CH_2-CH-CH_2-O-$$
$$\overset{|}{OH} \qquad\qquad\qquad \overset{|}{OH}\ \overset{|}{OH} \qquad\qquad\qquad \overset{|}{OH}$$

(I)   (II)   (III) is a result of imperfect dehydrochlorination or hydrolysis of glycidyl groups or of their reaction with free phenolic or hydroxy groups [19,20]. The presence of free phenolic groups is lěss probable. The high rate of anhydride hydrolysis [21] indicates that free carboxy groups may be present.

In the system epoxide (epoxy resin) — anhydride, we can thus expect the presence of anhydride, epoxy- and proton donor groups. In their study of the reaction mechanism, Fisch and Hofmann [20,22−24] proposed a sequence of reactions leading to the crosslinking of epoxy resins or to the formation of linear polyesters. The first step is the reaction of the anhydride with hydroxyl groups giving a monoester (Eq. (1))

$$R^1-OH \ + \ \overset{O}{\underset{C}{\parallel}}\diagdown^O\diagup\overset{O}{\underset{C}{\parallel}} \ \longrightarrow \ R^1-O-\overset{O}{\underset{C}{\parallel}}\diagdown\diagup\overset{O}{\underset{C}{\parallel}}-OH \tag{1}$$

The carboxy group formed in Eq. (1) reacts with the epoxide to yield a diester, and a secondary hydroxy group is formed.

$$R^1-O-\overset{O}{\underset{C}{\parallel}}\diagdown\diagup\overset{O}{\underset{C}{\parallel}}-OH \ + \ CH_2-CH-R^2 \longrightarrow R^1-O-\overset{O}{\underset{C}{\parallel}}\diagdown\diagup\overset{O}{\underset{C}{\parallel}}-O-CH_2-CH-OH \tag{2}$$

By repeating these reactions, either a linear polyester is obtained or crosslinking takes place. These reactions were confirmed by the fact that the amounts of mono- and diester produced are stoichiometrically equivalent to the amount of anhydride consumed and that the initial rate of production of monoesters is equal to the rate of consumption of anhydride and considerably higher than the rate of diester formation. An alternative is the reaction of free acid with epoxide:

$$R^3-COOH \ + \ H_2C-CH-R^2 \longrightarrow R^3-COO-CH_2-CH-OH \tag{3}$$

which accelerates the curing reaction of the system epoxy resin anhydride [25].

On the other hand, the consumption of epoxide in the non-catalyzed reaction is higher than that of anhydride, corresponding to 0.5–0.85 mol anhydride/mol epoxide [19,20,22−24,26,27]. Fisch and Hofmann [20,22−24] explained this difference by the reaction of secondary hydroxy group with the epoxide, the secondary hydroxy being reformed.

$$R^4-OH \ + \ H_2C-CH-R^2 \longrightarrow R^4-O-CH_2-CH-OH \tag{4}$$

This reaction is in fact the homopolymerization of the epoxide. The reactivity of the secondary hydroxy group with respect to the epoxy group is intermediate between that of tertiary and primary hydroxy group [26].

Shimazaki and Kozima [28,29] confirmed reactions (1), (2) and (4) by IR spectroscopy and showed that the reaction (4) is responsible for about 30 % loss of epoxides.

Another reaction is caused by moisture and involves hydrolysis of the epoxide [21,26].

$$H_2O \ + \ H_2C{-}CH{-}R^2 \ \longrightarrow \ HO{-}CH_2{-}CH{-}OH \qquad\qquad (5)$$
$$\underset{O}{\diagdown\diagup} \qquad\qquad\qquad\qquad \underset{R^2}{|}$$

However, hydrolysis of the anhydride, if present, will predominate.

Reactions (1) and (3) could be considered as initiation by proton-donor compounds present in the reaction system. The data reported by Kannebley [30] and Arnold [19] (Table 1) show that the rate of reaction (1) is higher than that of process (2); thus, solvolysis of the anhydride is faster than the formation of the diester. This, of course, does not mean that reaction (3) proceeds more slowly than reaction (1). Because of the lower pK_a value of the dissociation of the first carboxy group in dicarboxylic acids with respect to the second one, reaction (3) may be assumed to be faster than reaction (2). Moisture causes hydrolysis of the anhydride [21] and initiation occurs by reaction (3). The presence of secondary hydroxy groups in epoxy resins increases the probability of reaction (1), i.e. formation of the monoester. If we do not consider primary stages of the reaction, where moisture and free acids are consumed, curing or chain growth in the absence of catalysts are determined by reactions (1) and (2). In the absence of a catalyst the alcohol-glycidyl ether reaction (Eq. (4)) proceeds rather slowly; a temperature of 200 °C or higher was required to obtain measurable reaction rates [26,33]. Therefore, Fisch et al. [23] assume etherification to be catalyzed by the anhydride or by free carboxy groups. Nevertheless, the study of the kinetics of epoxy resin curing [34–36] revealed that no etherification occurs. The assumption was then made that homopolymerization of epoxides takes place only after gel formation [34–36], probably as a result of the local concentration changes of the reacting components.

The consumption of epoxide and anhydride is the same during curing [36]. The analysis of the conversion curves showed that curing reaction is second order with respect to epoxide and anhydride concentrations, i.e. first order with respect to each reagent,

$$-\frac{d\,[\text{epoxide}]}{dt} = -\frac{d\,[\text{anhydride}]}{dt} = k_a\,[\text{epoxide}] \cdot [\text{anhydride}] \qquad (6)$$

where k_a is the observed second-order rate constant. In the presence of proton-donor compounds, k_a also includes their effect in terms of Eq. (7):

$$k_a = k\,[\text{proton-donor}] \qquad\qquad (7)$$

In this case, the reaction rate is third order whereas at constant concentration of proton donors in the system, it is pseudosecond order. The gel time t_g is then inversely proportional to the proton donor concentration [36].

Kannebley [30] and Sorokin et al. [31,32] obtained second-order rate constants for the relevant model bimolecular reactions whereas Arnold [19] and Doszlop et al. [37] consider curing of epoxy resin with anhydride or reactions of monoepoxides with various proton-donor compounds to be first order (Table 1). Considering the mechanism of individual reactions which proceed during curing or in the non-catalyzed

Table 1. Kinetic parameters for non-catalyzed epoxide-anhydride-proton donor reactions[a]

Epoxide	An-hydride	Proton donor	Equa-tion	Overall reaction order	$10^5 \times k$[b]	T (°C)	E_a kJ/mol	Ref.
Araldite 6060	Harcure A	sec-OH	1	1	17	122		19)
Araldite 6060	Harcure A	sec-OH	2	1	37	122		19)
	MA	CHX	1	2	3.9	60.4	} 48.2	30)
	MA	CHX	1	2	6.2	70		30)
	MA	CHX	1	2	9.75	80.5		30)
	PA	CHX	1	2	5.1	60	} 44.9	30)
	PA	CHX	1	2	13	80		30)
	PA	CHX	1	2	28	100		30)
	HHPA	CHX	1	2	2.9	60	} 45.4	30)
	HHPA	CHX	1	2	7.3	80.2		30)
	HHPA	CHX	1	2	17	100		30)
DGEBA		BE-HET	2	2	9.8	70	} 49.0	30)
DGEBA		BE-HET	2	2	16	80		30)
DGEBA		BE-HET	2	2	24	90		30)
DGEBA		BE-HET	2	2	40	100		30)
DGEBA		BE-MA	2	2	2.4	80	} 54.8	30)
DGEBA		BE-MA	2	2	4.3	90		30)
DGEBA		BE-MA	2	2	7.2	100		30)
DGEBA		BE-PA	2	2	1.8	80	} 59.0	30)
DGEBA		BE-PA	2	2	3.2	90		30)
DGEBA		BE-PA	2	2	5.6	100		30)
DGEBA		BE-SA	2	2	2.9	105	} 68.2	30)
DGEBA		BE-SA	2	2	3.8	110		30)
DGEBA		BE-SA	2	2	4.6	114.5		30)
DGEBA		BE-SA	2	2	6.4	120		30)
DGEBA		BE-HHPA	2	2	0.9	100	} 74.5	30)
DGEBA		BE-HHPA	2	2	1.7	110		30)
DGEBA		BE-HHPA	2	2	2.9	120		30)
DGEBA		BE-HHPA	2	2	5.2	130		30)
	PA	butanol	1	2	11.3	100		31,32)
PGE		BuE-PA	2	2	0.04	100		31,32)
Cardura E		PLA	3	1	11.0	100	} 84.6	37)
Cardura E		PLA	3	1	23.2	110		37)
Cardura E		PLA	3	1	43.7	120		37)
Cardura E		BA	3	1	36.3	90	} 65.7	37)
Cardura E		BA	3	1	44.5	100		37)
Cardura E		BA	3	1	96.2	110		37)
Cardura E		n-octanol	4	1	2.05	150	} 45.6	37)
Cardura E		n-octanol	4	1	2.81	158.5		37)
Cardura E		n-octanol	4	1	3.01	168		37)
Cardura E		i-octanol	4	1	1.58	150		37)

[a] For symbols cf. List of abbreviations
[b] Rate constants in s^{-1} for first-order and in $l \cdot mol^{-1} \cdot s^{-1}$ for second-order reaction

formation of polyesters proposed by Fisch and Hofmann [20, 22-24], we assume second-order bimolecular elementary reactions.

Tanaka and Kakiuchi [35, 36] proposed a new mechanism of non-catalyzed copolymerization of epoxides with anhydrides. In the presence of proton donors, they expected the formation of a transition ternary complex composed of all three components of the reaction system. The proposed mechanism (Eqs. (8–10)) is similar to the reaction of phenol with epoxide catalyzed by phenolate [38].

$$R^1OH \; + \; H_2C{-}CH{-}R^2 \; \rightleftharpoons \; R^1{-}O{-}H\ldots O \hspace{-2pt}\underset{CH{-}R^2}{\overset{CH_2}{<}} \tag{8}$$

where R^1 is alkyl for alcohols, phenyl for phenols, and acyl for acids.

$$\text{(9)}$$

where R^1 is either alkyl or phenyl

$$\text{(10)}$$

In the first step (Eq. (8)), a binary complex of the proton donor and epoxy oxygen is formed. On addition of the anhydride, a ternary transition complex is formed which decomposes to a diester bearing a secondary hydroxy group (Eq. (9)). If the proton donor is a carboxy group, a monoester is formed and the carboxy group is regenerated (Eq. (10)). This scheme shows that, apart from phenol, the original initiating group is renewed in the growth reaction, i.e. the original concentration and distribution of hydroxy and carboxy groups might be retained in the reaction mixture, which has not been confirmed by chemical analysis of the reaction mixtures. Both in the reaction of epoxy resins [20] and in the model reaction between epoxide and anhydride in the presence of alcohols [23], the concentration of the hydroxy groups decreases in the first step and reaches a certain equilibrium value. The formation of transition ternary complexes held together only by polar interactions is very improbable. The copolyaddition reactions according to Eqs. (1)–(3), involving proton donors, epoxides and anhydrides seem to be more probable.

3 Anionic Copolymerization

In contrast to the non-catalyzed reaction, the base-initiated copolymerization was found to be a specific reaction [35,36,39–45] and the consumption of both monomers, epoxide and anhydride, is the same. The initiator not only affects the rate of copolymerization but also suppresses the undesirable homopolymerization of the epoxide. At equimolar ratio, epoxide and anhydride are strictly bifunctional.

Anionic initiators for copolymerization of epoxides can be divided into two groups. The first group comprises salts of inorganic and organic acids and the second group Lewis bases. Among them, tertiary amines are the most often used. Both types of initiators for the polyreaction of epoxides with anhydrides have some properties in common:
a) considerable acceleration of copolymerization,
b) anionic mechanism of copolymerization,
c) alternating composition of the copolymer formed,
d) suppression of side reactions of homopolymerization of epoxides.

3.1 Reactivity of Cyclic Ethers and Anhydrides

Under the same conditions, the reactivity of three-membered cyclic ethers in anionic copolymerization with cyclic anhydrides is higher than that of four-membered ethers [41]. Higher membered cyclic ethers can polymerize or copolymerize with anhydrides only by a cationic mechanism [42,46], whereby not only alternating copolymer but also a great number of polyether sequences are formed. This difference in reactivity is evidently associated with the basicity of cyclic ethers, three-membered ethers having the lowest basicity [47,48]. The lower basicity causes a lower reactivity of the epoxide (cyclic ether) in competitive reactions or in copolymerization with other cyclic monomers compared with the expected reactivity which follows from the strain in the ring. The strain energy, taken as the difference between the experimental and calculated heats of formation was found to be 54.4 kJ/mol for ethylene oxide [49].

The subsequent reactivity of substituted epoxides with respect to ethanol (Table 2) for ionic mechanisms has been reported [50]. Table 2 shows that the relative reaction rates between epoxide and ethanol are in the reverse order for acid- and base-catalyzed reactions respectively, and the polar effect of substituents is weaker for base catalysis than for acid catalysis. The less basic three-membered ethers are, due to their higher acidity, more sensitive to nucleophilic attack and are subjected to anionic polymerization and copolymerization. The effect of nucleophilic reagents decreases as the basicity and the size of the ring increase. This agrees with general conclusions inferred from papers of Parker and Isaacs [50, 51]: electron-withdrawing groups

Table 2. Relative rate constants for the reaction of epoxides with ethanol at 50 °C. [47] (Reproduced by courtesy of Marcel Dekker, Inc.)

Epoxide	Catalyst	
	$HClO_4$	C_2H_5ONa
Epichlorohydrin	0.065	4.8
Glycidol	0.41	2.0
Ethylene Oxide	(1)	(1)
Propylene Oxide	54.0	0.50
trans-2,3-Epoxybutane	119	0.03
cis-2,3-Epoxybutane	238	0.06
Isobutylene Oxide	5100	0.06
Styrene Oxide	$> 10^4$	0.14

increase the rate of reaction with nucleophilic reagents and decrease the reaction rate with electrophilic compounds. Tanaka and Kakiuchi [52] studied the polar effect of substituents for derivatives of phenylglycidyl ether in the copolymerization with hexahydrophthalic anhydride in solvents of different polarity. They obtained linear correlations between the logarithms of the rate constants and Hammett σ_p constants (Table 3). Reaction parameters ϱ are positive for both the initiation and the stationary regions; therefore, electron-withdrawing substituents increase the rate of copolymerization, copolymerization being of the S_N^2 type. Phenylglycidyl ethers with electron-donating substituents have lower rate constants and higher activation energies, enthalpies and frequency factors [56].

Table 3. Hammett reaction constants ϱ for the copolymerization of p-NO$_2$, p-CO$_2$CH$_3$, p-Cl, p-H, p-CH$_3$, and p-OCH$_3$ substituted phenylglycidyl ethers with hexahydrophthalic anhydride initiated with tri-n-butylamine[a]

Rate constant	Temperature (°C)[b]			Solvent[c]		
	94	103	118	$NO_2C_6H_5$	o-Cl$_2$C$_6$H$_4$	ClC$_6$H$_5$
Initial second-order	0.31	0.28	0.26	0.31	0.29	0.29
Maximum first-order	0.26	0.27	0.27	0.32	0.30	0.28

[a] Data of Tanaka and Kakiuchi [52];
[b] In mixture of xylenes;
[c] At 103 °C

Little is known about the reactivity of anhydrides in copolymerizations connected with ring opening. Tsirkin et al. [53] have reported a decrease in the reactivity of anhydrides in anionic copolymerization with epoxides with increasing gelation time: maleic > phthalic > tetrahydrophthalic > methyltetrahydrophthalic anhydrides.

Another reactivity order is reported for the copolymerization of anhydrides with phenylglycidyl ether and with 3,3-bis-chloromethyloxacyclobutane [41]:
glutaric > phthalic > succinic anhydrides.

These data show that the reactivity of anhydrides rises with increasing unsaturation of the anhydride ring and thus with decreasing basicity of the anhydride oxygen. Electron-withdrawing substituents should thus increase the reactivity of anhydrides in anionic copolymerization, similarly to epoxides.

3.2 Initiation by Inorganic and Organic Salts

Hamann and co-workers [41,42,54,55] studied the copolymerization of epoxides with cyclic anhydrides initiated by inorganic and organic salts. For the preparation of polyesters they used phenylglycidyl ether, cyclic carbonates or sulfites of aliphatic diols. At 200 °C, SO_2 or CO_2 split off from these compounds and epoxides are formed in situ. Copolymerization of epichlorohydrine with phthalic anhydride in solution or in melt was performed by Šňupárek and Mleziva [43] who, in a similar manner as Hamann [41], found the same rate of monomer consumption during polymerization. In order to confirm the formation of alternating copolyester, Hilt et al. [54,55] performed a hydrolysis of the copolymer. Thin-layer and gas chromatography showed that the content of polyether sequences in the hydrolysis products was below the sensitivity limit of the method (less than 0.1 mol-%), not only for equimolar initial mixtures of comonomers but also for an excess of anhydride. With an excess of epoxide, however, a shoulder appears on the gas chromatogram of the hydrolyzates, corresponding to the presence of diethylene glycol at low initiator concentration and at a relatively short reaction time. Therefore, the content of polyether can be assumed to be higher at higher epoxide conversion. From the changes in UV spectra recorded during copolymerization of 2-hydroxy-4-(2,3-epoxypropoxy) benzophenone with phthalic anhydride [7] it follows that at equimolar ratio of monomers or with an excess of anhydride an alternating copolymer is formed and that the reaction is highly specific. With a molar excess of epoxide, after the anhydride has been consumed for the formation of polyester units, polyetherification

3.2.1 Character of Copolymerization and Initiator Effect

Hilt et al. [42,54] found by conductivity measurements that the copolymerization of epoxides with cyclic anhydrides initiated by alkali salts is of ionic character. Lustoň and Maňásek [56,57], who used ammonium salts, came to the same conclusion. The rate of copolymerization increases linearly with rising initiator concentration (Fig. 1, the slope of the curve log k vs · log c_{In} is unity) but only up to a limit which depends on the solubility of the initiator in the reaction system [54]. A rise in the copolymerization rate is accompanied by an increase in the conductivity of the reaction

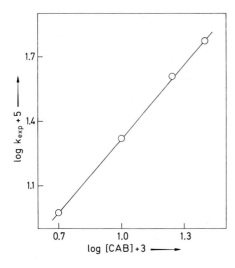

Fig. 1. Effect of hexadecyltrimethylammonium bromide (CAB) on the copolymerization of 2-hydroxy-4-(2,3-epoxypropoxy)benzophenone (0.5 mol/l) with phthalic anhydride (0.5 mol/l) in nitrobenzene at 120 °C [56]

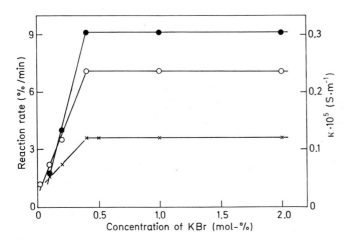

Fig. 2. Effect of KBr on the copolymerization of equimolar mixtures of ethylene glycol carbonate with phthalic anhydrite. [54] (Reproduced by courtesy of Hüthig and Wepf Verlag). 1 — conductivity at 120 °C; 2 — reaction rate at 200 °C; 3 — reaction rate at 180 °C

system which reaches the limiting value, depending on the type and concentration of the initiator and nature of solvent (Fig. 2). The results indicate that initiation does not occur on the surface of the initiator as has been expected [41] but in solution [54,56,57]. In the initiation by ammonium salts [57], the conductivity of the copolymerization mixture was not constant during experiment but decreased after passing through a maximum (Fig. 3). The decrease in conductivity is interpreted by a lowering of the mobility of ionic particles in solution resulting from the increase in molecular weight of the growing polymer and the rise of the viscosity of the system.

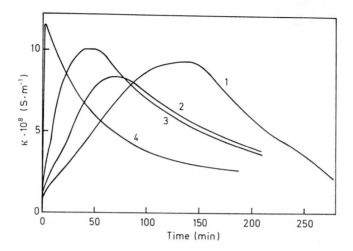

Fig. 3. Conductivity curves of the copolymerization of 2-hydroxy-4-(2,3-epoxypropoxy)benzophenone (0.5 mol/l) with phthalic anhydride (0.5 mol/l) in o-xylene at 120 °C initiated by various initiators. [57] 1 — tri-n-hexylamine (0.025 mol/l); 2 — tri-n-hexylamine (0.025 mol/l) and cyclohexanol (0.025 mol je l); 3 — tri-n-hexylamine (0.025 mol/l) and benzoic acid (0.025 mol/l); 4 — hexadecyltrimethyl-ammonium bromide (0.005 mol/l)

For alkali metal chlorides, the initiation efficiency increases in the order:

$$Li^+ < Na^+ < K^+ < Rb^+ < Cs^+ \ ^{54)}$$

and for ammonium bromides it decreases in the order:

tetrabutyl > hexadecyltrimethyl > hexadecylpyridinium > tetramethyl .

For iodides, the order of efficiency is: tetrabutyl > tetraethyl > tetramethyl [56]. A rise in the catalytic effect of initiator is observed for alkali salts and for ammonium salts with increasing diameter of the cation, hence with growing distance between charge centres. This is in agreement with an increase of the electropositivity of cations and the increasing ability of salts to dissociate [54]. Despite of this, the remarkable efficiency of lithium salts in curing of epoxy resins [58] or in copolymerization reactions [41] was confirmed in some papers.

The effect of the anion on the copolymerization rate is controversial. Hilt et al. [41] established for different sodium halides the following order of efficiency: $F^- < Cl^- < Br^- < J^-$. This order was interpreted on the basis of the increase in nucleophilicity or polarizability of the anions. Šňupárek and Mleziva [43] observed that the reaction rate of the benzoate anion was about twice as high as that of the bromide anion in the copolymerization of epichlorohydrine with phthalic anhydride. On the other hand, Lustoň and Maňásek [56] did not detect any effect of the anion size on the copolymerization rate initiated by tetramethylammonium and tetrabutylammonium halides.

The first step of the mechanism of initiation and copolymerization (cf. 3.2.3) is the reaction between the epoxide and the initiator (Eq. (11)).

$$A^- \, Cat^+ \; + \; H_2C\!-\!CH\!-\!R \; \longrightarrow \; A\!-\!CH_2\!-\!\underset{R}{CH}\!-\!O^- \, Cat^+ \qquad (11)$$

The course of this reaction was proved by Hilt et al. [42,54] who used as copolymerization initiators ^{14}C-labelled sodium and potassium benzoate. The activity of the prepared copolymer is due to the labelled and chemically bound initiator anion. This reaction is analogous to the analytic determination of epoxides by hydrogen halides [59] but instead of inactive halogen hydrine generaled according to Eq. (12), an ionic particle capable of initiating copolymerization is formed.

$$A^- \; ^+H \; + \; CH_2\!-\!CH\!-\!R \; \longrightarrow \; A\!-\!CH_2\!-\!\underset{R}{CH}\!-\!OH \qquad (12)$$

From the analogy with epoxide determination, we can assume that, despite the less polar medium in copolymerization, the rate of reaction (11) is high at elevated temperatures and that the reaction proceeds irreversibly. Therefore, pyridinium perchlorate, an ammonium salt with protonized nitrogen, does not initiate copolymerization but addition to the epoxy group occurs [56].

If the initiator is consumed rapidly and irreversibly in the first stages of copolymerization, the anions formed, irrespective of whether alkoxide or carboxylate, are the same for the given copolymerization system. Taking into account the dissociation effect of the initiator and the growing chain, the copolymerizing system can be illustrated as follows [60,61]:

$$A - B \rightleftarrows A^+ {}^-B \rightleftarrows \; A^+/solvent/B^- \qquad \rightleftarrows A^+/solvent + B^-/solvent \qquad (13)$$

| Contact ion pair | Solvent — separated ion pair | Completely dissociated ion pair-free ions |

The solution thus consists of different particles denoted as contact ion pairs, solvent-separated ion pairs and free ions. The fraction of the individual particles depends on the type of salt, type of solvent, polymerization system, temperature, and salt concentration. The catalytic effect of these particles may be very different as is evident in anionic polymerization of vinyl monomers. For instance, free polystyryl anion is 800times more reactive than its ion pair with the sodium counterion [60]. From this fact it follows that, although the portion of free ions is small in the reaction system, they may play an important role. On the other hand, anionic polymerization and copolymerization of heterocycles proceeds mostly via ion pairs. This is due to a strong localization of the negative charge on the chain-end heteroatom which strongly stabilizes the ion pair itself [62]. Ionic dissociation constants and ion contributions to the reaction kinetics are usually low. This means that for heterocycles the difference between the catalytic effect of ion pairs and free ions is much weaker than for the polymerization of unsaturated compounds. This is well documented by the copolymerization of anhydrides with epoxides where the substitution of o-xylene by nitrobenzene solvent causes a rise in the copolymerization rate by about 50 % while the conductivity in nitrobenzene is by two orders of magnitude higher [56]. Copolymerization is thus initiated by ion pairs as well as by free ions, the difference in the efficiency of both active particles being not great. This is supported by the values of activation energies for the copolymerization of phenyl-glycidyl ether with phthalic anhydride initiated by sodium(II) phthalate (72.4 kJ/mol) [41] or for the copolymerization of 2-hydroxy-4-(2,3-epoxypropoxy)benzophenone with phthalic anhydride catalyzed by hexadecyltrimethylammonium bromide in nitrobenzene (73.5 kJ/mol) [56] which are much higher than in anionic styrene poly-merization (21–25 kJ/mol) [60]; the frequency factor is lower (log A = 6.64 [56], log A = 8–9 [60]). A "normal" enthalpy-entropy effect is observed in copolymerization whereas a negative temperature effect on dissociation of ion pairs (i.e. formation of free ions with decreasing temperature [60]), which can give even negative values for apparent activation energies of anionic polymerization of vinyl monomers is not evident. This supports the assumption of a smaller difference between the activity of free ions and ion pairs. However, the rate constants for particular active centers are not available for any copolymerization system.

3.2.2 Solvent Effect

Solvent polarity influences the rate of copolymerization. Thus with increasing dielectric constant of the solvent, the copolymerization rate rises as a result of the increase in the dissociation constants of the active species. The apparent rate constant for the copolymerization of 2-hydroxy-4-(2,3-epoxypropoxy)benzophenone with phthalic anhydride, initiated by hexadecyltrimethylammonium bromide [56], increases from 4.65×10^{-4} s^{-1} in o-xylene to 6.84×10^{-4} s^{-1} in nitrobenzene. Hilt et al. [54] proposed a suitable model illustrating the effect of solvent polarity in the copoly-merization of phthalic anhydride with ethylene glycol carbonate in a mixture of nitrobenzene and trichlorobenzene (Table 4). With increasing fraction of the more polar nitrobenzene, the rate of copolymerization increases.

Not only the solvent contributes to the dieletric constant of the copolymerization mixture (Table 4) but also the monomers. With respect to the formation of the

Table 4. Reaction rates and dielectric constants of solvents and reaction mixtures for the copoly-merization of ethylene glycol carbonate (0.1 mol) with phthalic anhydride (0.1 mol) in 100 ml of solvent initiated with KCl (0.001 mol-%) at 200 °C. [54] (Reproduced by courtesy of Hüthig and Wepf Verlag.)

Solvent	Dielectric constant at 200 °C		Reaction rate % conversion/min
	solvent	solution	
Nitrobenzene (I)	16.6	20.9	1.15
Mixture of 3 parts of I and 1 part of II	12.1	16.7	0.94
Mixture of 2 parts of I and 2 parts of II	8.8	13.3	0.51
Mixture of 1 part of I and 3 parts of II	5.6	10.4	0.26
Trichlorobenzene (II)	3.4	7.2	0.11

alternating copolymer, the accelerating solvent effect is not connected with a preferential solvation of monomers but results from the increased dissociation of ion pairs in the more polar medium.

3.2.3 Mechanism of Copolymerization

The first scheme of the reaction between acetic anhydride and phenylglycidyl ether initiated by potassium acetate was proposed by Schechter and Wynstra [26] (Eqs. (14) and (15)).

$$CH_3COO^- + H_2C\!-\!CH\!-\!R \longrightarrow CH_3COO\!-\!CH_2\!-\!CH\!-\!O^- \qquad (14)$$
$$\overset{\diagdown\!\diagup}{O} \qquad\qquad \overset{|}{R}$$

$$CH_3COO\!-\!CH_2\!-\!\underset{\underset{R}{|}}{CH}\!-\!O^- + (CH_3CO)_2O \longrightarrow$$

$$CH_3COO\!-\!CH_2\!-\!\underset{\underset{R}{|}}{CH}\!-\!OOCCH_3 + CH_3COO^- \qquad (15)$$

The acetate anion opens an oxirane ring giving an alkoxide anion (Eq. (14)) which then reacts with the anhydride to yield a diester and a carboxylate anion (Eq. (15)).

Hamann and co-workers [41,42,54] applied this mechanism to the copolymerization of epoxides with cyclic anhydrides initiated by organic and inorganic salts. They suggested individual stages of copolymerization according to Eqs. (16–23).[1]

Initiator dissociation:

$$KX \rightleftarrows K^{+-}X \rightleftarrows K^+ + X^- \qquad (16)$$

Initiation:

$$X^- + H_2C\!-\!CH\!-\!R \longrightarrow X\!-\!CH_2\!-\!\underset{\underset{R}{|}}{CH}\!-\!O^- \qquad (17a)$$
$$\overset{\diagdown\!\diagup}{O}$$

$$X^- + H_2C\!-\!\!-\!\!-\!CH_2 \longrightarrow X\!-\!CH_2\!-\!CH_2\!-\!O^- + ZO_2 \qquad (17b)$$
$$\underset{\underset{\underset{O}{\|}}{Z}}{\underset{|}{O}\qquad\underset{|}{O}}$$

where Z is C or S
Propagation:

$$X\!-\!CH_2\!-\!\underset{\underset{R}{|}}{CH}\!-\!O^- + \overset{O}{\underset{}{\overset{\|}{C}}}\!\diagdown O \diagup\!\overset{O}{\overset{\|}{C}} \longrightarrow X\!-\!CH_2\!-\!\underset{\underset{R}{|}}{CH}\!-\!O\!-\!\overset{O}{\overset{\|}{C}}\qquad\overset{O}{\overset{\|}{C}}\!-\!O^- \qquad (18)$$

---

[1] Glycol carbonates and glycol sulfites are also included; they split off respective oxides and form epoxides in situ during copolymerization [54]

$$\text{R}^1-\overset{\overset{\displaystyle O}{\|}}{\text{C}}-\text{O}^- \; + \; \underset{\underset{\displaystyle O}{\diagup\diagdown}}{\text{H}_2\text{C}-\text{CH}-\text{R}} \longrightarrow \; \text{R}^1-\overset{\overset{\displaystyle O}{\|}}{\text{C}}-\text{O}-\text{CH}_2-\underset{\underset{\displaystyle R}{|}}{\text{CH}}-\text{O}^- \qquad (19)$$

$$\text{R}^1-\overset{\overset{\displaystyle O}{\|}}{\text{C}}-\text{O}^- \; + \; \underset{\underset{\underset{\displaystyle O}{\overset{\displaystyle \|}{Z}}}{O\diagdown\diagup O}}{\text{H}_2\text{C}\text{-----}\text{CH}_2} \longrightarrow \; \text{R}^1-\overset{\overset{\displaystyle O}{\|}}{\text{C}}-\text{O}-\text{CH}_2-\text{CH}_2-\text{O}^- \; + \; \text{ZO}_2 \qquad (20)$$

Termination and transfer:

$$\text{P}-\text{CH}_2-\underset{\underset{\displaystyle R}{|}}{\text{CH}}-\text{O}^- \; + \; \text{HX} \longrightarrow \text{P}-\text{CH}_2-\underset{\underset{\displaystyle R}{|}}{\text{CH}}-\text{OH} \; + \; \text{X}^- \qquad (21)$$

$$\text{P}-\overset{\overset{\displaystyle O}{\|}}{\text{C}}-\text{O}^- \; + \; \text{HX} \longrightarrow \text{P}-\overset{\overset{\displaystyle O}{\|}}{\text{C}}-\text{OH} \; + \; \text{X}^- \qquad (22)$$

$$\text{P}-\overset{\overset{\displaystyle O}{\|}}{\text{C}}-\text{O}^- \; + \; \text{K}^+ \longrightarrow \text{P}-\overset{\overset{\displaystyle O}{\|}}{\text{C}}-\text{OK} \qquad (23)$$

According to this scheme, the initiation of copolymerization is determined by the dissociation of the initiator in the reaction medium (Eq. (16) (cf. 3.2.1) giving anions which initiate chain growth by the formation of primary active centres in the reaction with epoxides (Eq. (17a)) or with chemically similar compounds (Eq. (17b)) accompanied by ring opening. The initiator anion is irreversibly bound to the monomer at the start of the chain and the active centre is an alkoxide anion. Propagation proceeds through the reaction of the alkoxide anion with the anhydride. The carboxylate anion formed (Eq. (18)) reacts in the next propagation step with the cyclic ether (Eq. (19)), glycol carbonate or glycol sulfite (Eq. (20)) yielding again an alkoxide anion. Termination occurs by reaction of the propagating carboxylate or alkoxide anion with the electrophilic agent (proton from acid, glycol or water) which may be present in the reaction mixture (Eqs. (21) and (22)). Another termination involves recombination of the growing chain end with a cation according to Eq. (23) and a non-dissociated salt is formed again [42, 54]. Polyesters with this end group may be considered as living polymers, which, on addition of monomers, initiate again the copolymerization reaction [54]. The formation of an alcoholate at an excess of carboxy groups is not expected. As is seen from Eqs. (21) and (22), termination of the growing chain end is connected with the formation of a new anion able to initiate the growth of a new chain. As a matter of fact, this is a chain transfer. The hydroxy and carboxy groups formed can also take part in transfer and termination reactions.

The course of initiation was proved by the activity of the polyester prepared by using a labelled initiator [42, 54], it also follows from the analytical determination of the epoxide by halogenide salts [59]. The formation of alternating copolymer — polyester and thus propagation according to Eqs. (18)–(20) were confirmed by the same loss of epoxide and anhydride during copolymerization [41, 43] and by the analysis of hydrolyzed polyester [42, 54, 55] whereby formation of polyglycols was not proved. This shows that in basic initiation the reaction between alkoxide anion and epoxide (Eq. (24)) where polyether sequences are formed is suppressed.

$$P\text{--}CH_2\text{--}CH\text{--}O^- \;+\; CH_2\text{--}CH\text{--}R \;\xrightarrow{\;\times\;}\; P\text{--}CH_2\text{--}CH\text{--}O\text{--}CH_2\text{--}CH\text{--}O^- \qquad (24)$$

Termination and transfer reactions according to Eqs. (21) and (22) were proved by a decrease in the molecular weight of polymers in the presence of water, glycol and phthalic acid [42, 54]. The presence of potassium in a polyester the formation of which was initiated by labelled potassium benzoate indicates the course of the reaction (23).

Smith [63] has recently reported a new mechanism for the initiation by quaternary phosphonium salts. This mechanism assumes interaction of the acidic hydrogen atom of the alkyl group of phosphonium salts with epoxy oxygen as a nucleophile:

$$(25)$$

The epoxy-phosphonium salt adduct would then be vulnerable to attack by another epoxy molecule resulting in the formation of an oxonium ion (Eq. (26)) or it reacts with the anhydride to form a special monoester with hydroxy groups (Eq. (27)).

$$(26)$$

$$(27)$$

Smith [63] has also proposed another possibility, namely a direct reaction of the quaternary phosphonium salt with anhydride (Eq. (28)).

$$(28)$$

From the activation energy of 67 kJ/mol the author assumes that the dissociation of phosphonium salt-epoxide or phosphonium salt-anhydride complexes is the rate-determining step of initiation because cleavage of a hydrogen bond and transfer of a proton to the epoxy or anhydride oxygen are assumed.

According to this mechanism, at low concentrations, the initiator causes only the formation of "active hydrogen" (however, this "active hydrogen" is present in epoxy resins as primary or secondary hydroxy groups) and does not take part in propagation nor does it influence the propagation rate. According to Smith [63] phosphonium salts are involved in chain growth, only at higher initiator concentrations, e.g. directly by the formation of crosslinks.

However, it seems that this mechanism does not fully agree with experimental results. The rate of gel formation is higher in the curing of epoxy resin by phosphonium salts. Moreover, the activation energy is higher than the strength of hydrogen bonds and is close to the activation energies of 67 kJ/mol and 72.4 kJ/mol determined by Schwenk et al. [41] or 73.7 kJ/mol determined by Lustoň and Maňásek [56].

Therefore, a similarity between the mechanisms of copolymerization of epoxides with anhydrides initiated either by phosphonium salts or alkali metal or ammonium salts can be expected and copolymerization then proceeds according to Hamann's mechanism illustrated by Eqs. (16)–(23).

3.2.4 Kinetics of Copolymerization

Only few data are avalaible on the copolymerization of epoxides with cyclic anhydrides initiated by inorganic and organic salts.

Hilt et al. [42,54] reported that the conversion curves obtained from the melt are linear between 10 and 40% conversion and are thus of zero order with respect to

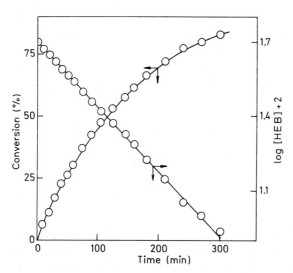

Fig. 4. Copolymerization of 2-hydroxy-4-(2,3-epoxypropoxy) benzophenone (HEB) (0.5 mol/l) with phthalic anhydride (0.5 mol/l) in nitrobenzene at 90 °C initiated by hexadecyltrimethylammonium bromide (0.025 mol/l) [56]

monomers. Recent results have, however, shown [56] that copolymerization is a first-order reaction both with respect to either monomer (Fig. 4) and to the initiator (Fig. 1) at equimolar ratio of monomers.

$$-\frac{d\,[\text{epoxide}]}{dt} = k_p\,[\text{initiator}]\,[\text{epoxide}] \tag{29}$$

where $k_p = \dfrac{k_2 k_3 K}{k_2 + k_3}$, k_2 and k_3 are the propagation rate constants (Eqs. (18)–(20)) and K is given by $K = K_d/(1 + K_d)$, K_d being the dissociation equilibrium constant of active centers in terms of Eqs. (13) or (16). At constant initiator concentration, copolymerization seems to be a first-order reaction.

For non-equimolar ratios of monomers with an excess of anhydride, the rate equation is

$$-\frac{d\,[\text{epoxide}]}{dt} = \frac{k_3 K\,[\text{initiator}]\,[\text{epoxide}]}{1 + k_3\,[\text{epoxide}]/k_2\,[\text{anhydride}]} \tag{30}$$

and, after modification and integration

$$\ln \frac{[\text{epoxide}]}{[\text{epoxide}]_0} = \frac{k_3}{k_2} \ln \frac{[\text{anhydride}]_0}{[\text{anhydride}]} - k_a t \tag{31}$$

where $k_a = (1 + k_3/k_2)\,k_p\,[\text{initiator}]$, the other symbols are the same as in Eq. (29).

Equation (31) allows the determination of the ratio of propagation rate constants k_3/k_2 by correlation of the experimental results obtained in copolymerization using equimolar ratios of monomers with the results obtained at non-equimolar monomer ratio and excess of anhydride (an excess of epoxide should not be used because of homopolymerization of the monomers at high degree of conversion). A value of k_3/k_2 equal to 0.2 ± 0.1 was found for the system 2-hydroxy-4-(2,3-epoxypropoxy) benzophenone-phthalic anhydride in nitrobenzene initiated by hexadecyltrimethyl-ammonium bromide [56].

This value shows that the rate of the reaction between alkoxide anion and anhydride (Eq. (18)) is about five times higher than that of the reaction of carboxylate anion with epoxide (Eq. (19)). Since it holds for the stationary state of copolymerization:

$$k_2\,[\text{anhydride}]\,[\text{alkoxide}] = k_3\,[\text{epoxide}]\,[\text{carboxylate}] \tag{32}$$

the value k_3/k_2 shows that the concentration of carboxylate active centres is higher than that of alkoxide anions. This also results from the much higher basicity of the alkoxide anion compared with the resonance — stabilized carboxylate anion. However, the absolute values of the rate constants for individual copolymerization steps have not yet been determined.

3.2.5 Molecular Weights

Data on molecular weights of polyesters can provide information on the mechanism and termination and transfer reactions. As follows from section 3.2.3 the co-polymerization of epoxides with cyclic anhydrides should proceed stoichiometrically without transfer or termination reactions, and the average degrees of copolymerization should only depend on the molar ratio of monomers to the initiator. Polymers with a narrow molecular weight distribution should be obtained.

Experimental results on the copolymerization of phthalic anhydride with ethylene glycol carbonate initiated by ^{14}C-labelled sodium benzoate (Table 5) or potassium benzoate show [41,42,54] that the molecular weights of the copolymers formed decrease with increasing initiator concentration.

Table 5. Concentration effect of ^{14}C-labelled sodium benzoate on the molecular weight of the polyester prepared from phthalic anhydride and ethylene glycol carbonate after 10 hours at 200 °C

Initiator conc. (mol-%)	M_v[a]	M_i[b]	M_c[c]	DP_c[d]	M_c/M_i
0.05	67000	399000	384000	2000	0.962
0.1	65000	184000	192000	1000	1.043
0.2	68000	102000	96000	500	0.941
0.4	62000	55900	48000	250	0.859
0.5	52000	55800	38400	200	0.688
0.714	53000	42400	26880	140	0.634
1.0	46000	27000	19200	100	0.711
2.2	29000	15400	8640	45	0.561
5.55	22000	7550	3456	18	0.458

[a] Viscosity average molecular weight [54];
[b] Molecular weight determined from the impulse number of the precipitated polyester according to

$$M_i = \frac{I \cdot M_{inic}}{P}$$

where I is the impulse number of the initiator (Imp/min · mg), M_{inic} the initiator molecular weight and P the impulse number of the polyester (Imp/min · mg) [54];
[c] Calculated molecular weight on the assumption of one initiator molecule per chain;
[d] Calculated degree of polymerization

Table 5 shows that, at low initiator concentrations, the molecular weights determined by viscometry (M_v) are lower than those calculated from the radio activity of the prepared polyester (M_i). As the initiator concentration increases, the relation becomes reverse. On the other hand, the molecular weights calculated on the assumption that one initiator molecule initiates the growth of one chain without considering transfer reactions (M_c) agree well with M_i for low initiator concentrations. With increasing initiator concentration, the ratio M_c/M_i decreases indicating that not all initiator molecules take part in initiation reactions. This results from the limited solubility of the initiator in a copolymerization mixture. We can simultaneously expect a low degree of transfer and termination reaction, i.e. a low content of proton donors.

In contrast to the copolymerization of cyclic carbonates, the molecular weights are lower in the epoxide copolymerization [6, 41, 43]. We assume that this is due to the presence of proton donors in the reaction mixture. They occur as impurities in epoxides [19, 20] or anhydrides, moisture in all components of the copolymerization system, or their presence is a consequence of the high rate of hydrolysis of cyclic anhydrides [21]. Proton donors added to the monomer feed remarkably decrease the molecular weight [42, 54], even in the copolymerization of ethylene glycol carbonate at 200 °C. Under these conditions, when recyclization of phthalic acid takes place [64] and the released CO_2 can tear off moisture to the gas phase, the molecular weight M_v decreases without proton donors from 45200 to 7100 in the presence of 5 % phthalic acid or ethylene glycol or to 9300 in the presence of 15 % water [42, 54].

Fractionation of polyesters, prepared from phthalic anhydride and ethylene glycol carbonate and initiated by sodium benzoate or by KCl (Fig. 5) revealed a broad molecular weight distribution [42]. This indicates a wide application of transfer reactions and participation of macromolecules in transfer reactions.

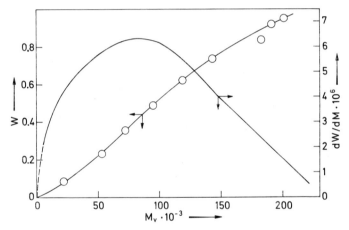

Fig. 5. Integral and differential distribution curves of polyesters [42] (Reproduced by courtesy of Hüthig and Wepf Verlag.). (Equimolar mixture of ethylene glycol carbonate and phthalic anhydride, 1 mol-% KCl, 15 h at 180 °C, and 24 h at 200 °C.)

The effect of the medium on the molecular weight of copolymers is evident in copolymerization. Schwenk et al. [41] obtained five to four times higher molecular weights in the melt than in copolymerization in aprotic solvents. In the copolymerization of glycidyl cinnamate with phthalic anhydride initiated by triethylbenzyl-ammonium chloride Nishikubo and coworkers [8] found higher viscosities in melt as compared with the copolymerization performed in solution. A different result was obtained only by Šňupárek and Mleziva [43] in the copolymerization of phthalic anhydride with epichlorohydrin. Nevertheless, the copolymerization of cyclic anhydrides with epoxides can yield polyesters with molecular weights higher than those obtained by polycondensation of dicarboxylic acids with diols without the application of low pressures and high temperatures.

3.3 Initiation by Lewis Bases

For the copolymerization of epoxides with cyclic anhydrides and curing of epoxy resins, Lewis bases such as tertiary amines are most frequently used as initiators. In this case, terminal epoxides react with cyclic anhydrides at equimolar ratios. The time dependence of the consumption of epoxide and anhydride is almost the same for curing [35,36] and for model copolymerizations [39,40,45]. The reaction is specific [39,40] to at least 99%. In contrast, the copolymerization with non-terminal epoxides does not exhibit this high specificity, probably because of steric hindrances. The copolymerization of vinylcyclohexene oxide or cyclohexene oxide is specific only to 75–80% and internal epoxides such as alkylepoxy stearates react with anhydrides only to 60–65%. On the other hand, in the reaction of epoxy resins with maleic anhydride the consumption of anhydride is faster [65], the products are discoloured and the gel is formed at a low anhydride conversion [39]. Fischer [39] assumes that the other resonance form of maleic anhydride is involved in the reaction according to Eq. (33).

$$
\begin{array}{l}
\mathrm{\sim\overset{O}{\overset{\|}{C}}-O^-} \ + \ \substack{^+CH-C\diagup^O\diagdown_O \\ | \\ CH=C\diagdown_{O^-}} \ \longrightarrow \ \mathrm{\sim\overset{O}{\overset{\|}{C}}-O-\underset{CH=C\diagdown_{O^-}}{\overset{|}{CH}}-C\diagup^O\diagdown_O} \ \rightleftharpoons
\end{array}
$$

$$
\rightleftharpoons \ \mathrm{\sim\overset{O}{\overset{\|}{C}}-O-\underset{^-CH-C\diagdown_O}{\overset{|}{CH}}-C\diagup^O\diagdown_O} \tag{33}
$$

Another explanation is the formation of a 1:1 salts of maleic anhydride with the tertiary amine (Eq. (34)) through generation of π- and σ-complexes. The resulting dark brown adduct exhibits a betaine-like structure [66].

$$
\begin{array}{ccccc}
\diagdown N: + & \substack{CH-C\diagup^O\diagdown_O \\ \| \\ CH-C\diagdown_O} & \longrightarrow & \substack{CH-C\diagup^O\diagdown_O \\ \| \\ CH-C\diagdown_O \\ \overset{\cdots}{N} \\ /|\diagdown} & \longrightarrow & \substack{CH-C\diagup^O\diagdown_O^{\delta-} \\ \| \\ CH-C\diagdown_O \\ \overset{\cdots}{N}^{\delta+} \\ /|\diagdown} & \longrightarrow
\end{array}
$$
$$
\begin{array}{ccc}
 & \pi\ \text{complex} & \sigma\ \text{complex}
\end{array}
$$

$$
\longrightarrow \ \substack{CH-\overset{O}{\overset{\|}{C}}-O^- \\ \| \\ CH-C-N^+\diagdown^< \\ \| \\ O} \tag{34}
$$

No similar effect was observed with other anhydrides.

3.3.1 Character of Copolymerization

For a long time, the copolymerization of epoxides with cyclic anhydrides initiated by tertiary amines has been assumed to proceed by an anionic mechanism. However, only recently was the presence of ions in a mixture of epoxide/anhydride/tertiary amine proved by conductivity measurements of the copolymerization solution [57]. Also the presence of carboxylate species in a copolymerization system was recently detected by IR-spectroscopy [67] as well as the formation of ammonium salts by ^{1}H-NMR spectroscopy [68]. The conductivity of the system was found to increase with time and to decrease gradually after reaching a maximum (Fig. 3). The time for reaching the maximum on the conductivity curve is equal to the induction period of copolymerization. The initial rise of the conductivity and induction period of copolymerization has been interpreted as being due to the increase in the concentration of active centres [52, 57], and a decrease in the conductivity is assumed to result from the growing molecular weight and rising solution viscosity [57]. However, the concentration of active centres remains constant [68]. At the same time, we can observe that in the presence of proton donors the maximum on the conductivity curve is reached sooner. Concerning the copolymerization mechanism this observation means that the stationary concentration of active centres is also reached earlier. Conductivities of the reaction mixtures are low and increase with rising polarity of the solvent. This increase indicates the dissociation of active centres and that only a part of ion pairs is dissociated in solution. Similar to the initiation by salts (cf. 3.2.1), copolymerization is initiated by both free ions and ion pairs.

3.3.2 Effect of Tertiary Amines

The structure of amines affects the rate of copolymerization. According to Fischer [39, 40], amines are weaker initiators than quaternary ammonium salts. For the curing of epoxy resins with hexahydrophthalic anhydride, Tanaka and Kakiuchi [35] reported for amines the following order of efficiency: N,N-dimethylbenzylamine > pyridine > trimethylamine > triethanolamine > N,N-dimethylaniline. No correlation has been found between the dissociation constants of amines and their initiation efficiency. Various amines were used by Arnold [19] for curing epoxy resins with anhydrides and by Lustoň and Maňásek [69] for model copolymerization of epoxides with anhydrides. Their results show that the copolymerization rate depends on the activity of the free electron pair on nitrogen. This activity is influenced by electronic and steric effects of the substituents of the tertiary amine. The efficiency of aliphatic tertiary amines decreases with rising alkyl chain length [69], i.e. with the decrease of the electronic effect and the increase in steric hindrances. Arnold [19] explained the difference between N,N-dimethylaniline and methylene-bis-(N,N-dimethylaniline), which initiate crosslinking, and aniline, N-methylaniline and methylenedianiline, which are not effective, by the electronic effects of the alkyl groups. A difference in the initiation efficiency between tertiary and secondary amines was also found by Lustoň and Maňásek [69]. Tertiary amines are efficient copolymerization initiators whereas in the case of secondary amines only addition reactions take place; there the reaction between secondary amines and anhydrides or epoxides can proceed. In the case of anhydride reactions, an amide group will be formed. However, there is no information available of the activity of an amide group to initiate the copolymerization reaction.

In the case of epoxide-amine reactions, β-hydroxyalkylamine should be generated where a hydrogen bond between the free electron pair of nitrogen and secondary hydroxy group can be formed (Eq. (35)):

$$R_2NH \;+\; \underset{O}{CH_2-CH-R} \;\longrightarrow\; \underset{\substack{|\\O-H}}{R-CH-CH_2} \overset{}{\underset{..\,NR_2}{\Big\backslash}} \tag{35}$$

This deactivation of the amino group in β-hydroxyalkylamines has already been proposed by Alvey [70] and is supported by the results of the initiation of copolymerization by quinoline and its 8-hydroxy derivative. Whereas a high initiation efficiency was found for quinoline, 8-hydroxyquinoline, a very effective chelating agent, does not initiate copolymerization [69].

Electron-withdrawing groups like phenyl lower the basicity of amines since the unpaired nitrogen electrons in arylamine are in resonance with the aromatic ring [19]. This accounts for the great difference in the reactivity of N,N-dimethylaniline and N,N-dimethylbenzylamine found by Tanaka and Kakiuchi [35]. The presence of an isolating group between the nitrogen atom and the aryl group prevents the resonance.

The initiation by aromatic tertiary amines is affected by the type of substituent and its position in the ring. A series of meta-substituted pyridines show a higher activity than ortho- or para-substituted derivatives with electron-releasing substituents. 3-Aminopyridine is much more reactive than 2-aminopyridine and 3-methylpyridine is more effective than 2- or 4-methylpyridine [19]. The strong electron-withdrawing substituent of 2-bromopyridine and its steric hindrance eliminate the effect of this amine. A 2,6-Dimethylpyridine was also found to display a steric effect. Thus, in the presence of this amine, the rate of copolymerization decreases by a factor of 4 compared with 3-methylpyridine [69].

For the characterization of the electronic and steric effects of substituents on the catalytic properties of tertiary aliphatic amines, Bogatkov et al. [71,72] proposed a correlation between the steric hindrance of $R^1R^2R^3N$ amines by means of the $E_N(E_N^0)$ constants which are identical with steric constants $E_s(E_s^0)$ according to Taft for isosteric hydrocarbon substrates $R^1R^2R^3C$. On this basis, they proposed the following linear correlation for the nucleophilic activity of amines:

$$\log k = \log k_0 + \varrho^* \Sigma \sigma^* + \delta E_N \tag{36}$$

Here, k and k_0 are the rate constants for the reaction initiated by a tertiary amine and a standard tertiary amine, respectively, ϱ^* is the reaction constant, σ^* are the substituent parameters, δ is the steric parameter of the reaction and $E_N(E_N^0)$ are steric parameters of the substituents (identical with the Taft's constants $E_s(E_s^0)$).

However, the experimental data in the literature do not allow to determine the effect of the structure of the amines on the rate of copolymerization of epoxides with cyclic anhydrides, nor to verify the validity of Eq. (36) for this type of reaction.

3.3.3 Mechanism of Copolymerization

Several mechanism have been proposed for the copolymerization of epoxides with cyclic anhydrides, initiated by tertiary amines. In one of the first papers on copolymerization mechanisms Fischer [39, 40] suggested the following scheme:

Initiation:

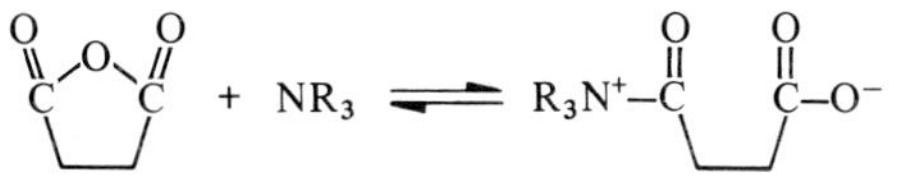

$$\text{(37)}$$

IV

Propagation:

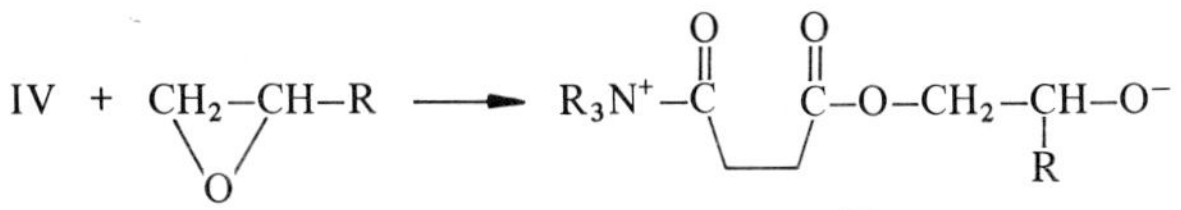

$$\text{(38)}$$

V

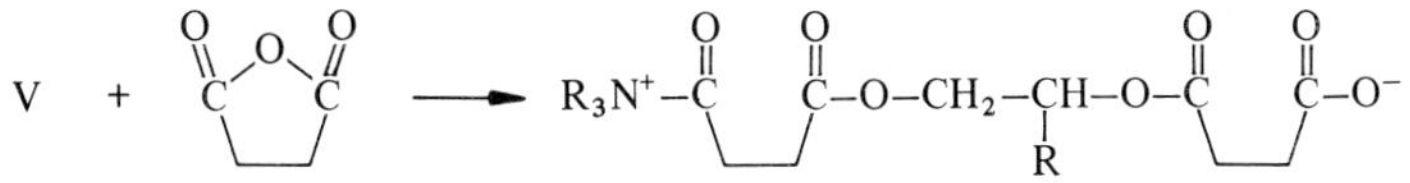

$$\text{(39)}$$

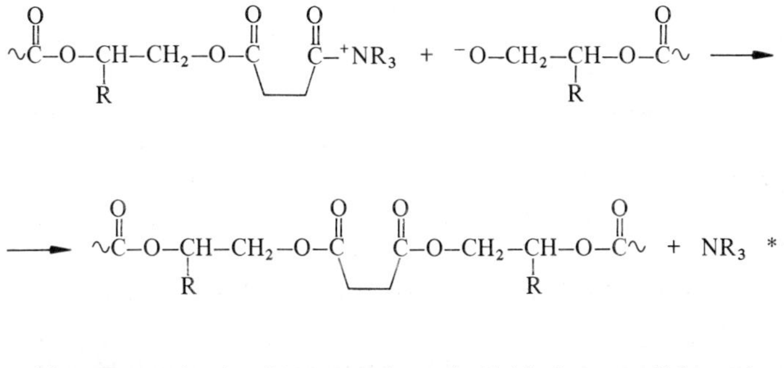

$$\text{(40)}$$

According to this scheme, in the first step the anhydride is reversibly activated by a tertiary amine with the formation of a betaine-like "Zwitterion" IV. An excess of the anhydride (in the presence of a catalytic amount of tertiary amine) causes a shift of the equilibrium to the betaine so that the portion of the backward reaction (Eq. (37)) is about 4%. In the first propagation step the zwitterion reacts with the epoxide to give compound V with a tertiary alkoxide residue. Compound V opens an anhydride ring in the next propagation step, yielding a carboxylate anion, and by alternation of these reactions, the chain grows and a polyester is formed. Chain growth is termed by reaction (40) in which a tertiary amine is regenerated.

---

[2] Compounds I–III are described on p. 3

This copolymerization scheme was modified by Tanaka and Kakiuchi [36)] who considered the case in which compounds with an active hydrogen atom HA are involved in copolymerization (Eqs. (41–44)).

$$R_3N + HA \rightleftharpoons [R_3N...HA] \tag{41}$$

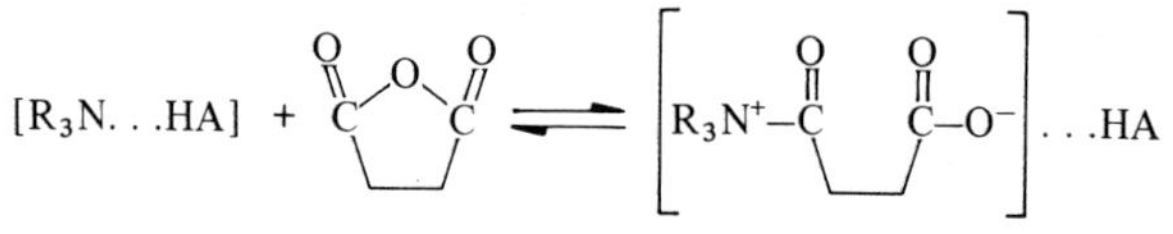

$$[R_3N...HA] + \text{(anhydride)} \rightleftharpoons \left[R_3N^+ - \overset{O}{\underset{\|}{C}} \quad \overset{O}{\underset{\|}{C}} - O^- \right]...HA \tag{42}$$

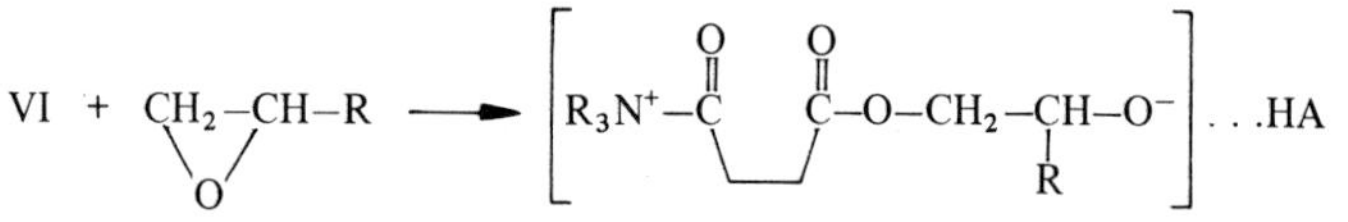

VI

$$VI + CH_2-CH-R \longrightarrow \left[R_3N^+ - \overset{O}{\underset{\|}{C}} \quad \overset{O}{\underset{\|}{C}} - O - CH_2 - \underset{R}{CH} - O^- \right]...HA \tag{43}$$

VII

$$VII + \text{(anhydride)} \longrightarrow \left[R_3N^+ - \overset{O}{\underset{\|}{C}} \quad \overset{O}{\underset{\|}{C}} - O - CH_2 - \underset{R}{CH} - O - \overset{O}{\underset{\|}{C}} \quad \overset{O}{\underset{\|}{C}} - O^- \right]...HA \tag{44}$$

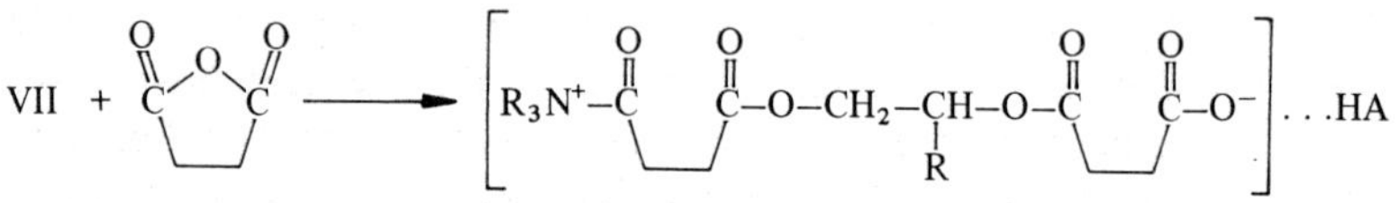

According to this mechanism, interaction between the tertiary amine and HA forms a hydrogen-bound associate, which then opens an anhydride ring giving a betaine-like structure with a carboxylate group. This carboxylate anion, which is considered to be the active centre of copolymerization, then reacts with the epoxide (Eq. (43)). The resulting alkoxide anion reacts with the anhydride (Eq. (44)). The anionic chain end is stabilized by HA. Chain growth occurs through repeated reactions (43) and (44).

Another mechanism of copolymerization was reported by Feltzin et al. [73)]. These authors also considered activation of tertiary amine by a compound with active hydrogen but with the formation of a quaternary ammonium base (Eq. (45)), being regarded as the real initiator of copolymerization. Initiation, propagation and termination reactions according to Feltzin are illustrated by schemes (46)–(49).

$$R_3N + HA \rightleftharpoons R_3HN^+ \ ^-A \tag{45}$$

Initiation:

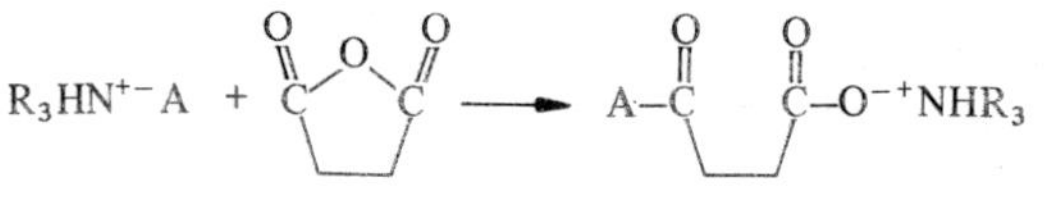

$$R_3HN^+ \ ^-A + \text{(anhydride)} \longrightarrow A - \overset{O}{\underset{\|}{C}} \quad \overset{O}{\underset{\|}{C}} - O^- \ ^+NHR_3 \tag{46}$$

Propagation:

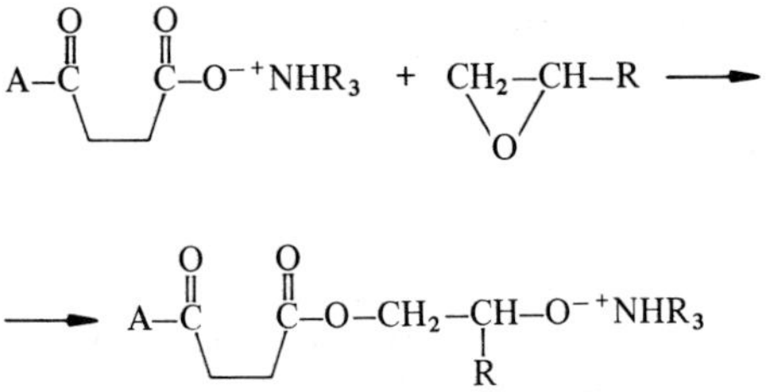

$$A-\overset{\overset{O}{\|}}{C}\qquad\overset{\overset{O}{\|}}{C}-O^-\,{}^+NHR_3 \;+\; \underset{\underset{O}{\diagdown\diagup}}{CH_2-CH-R} \longrightarrow$$

$$\longrightarrow A-\overset{\overset{O}{\|}}{C}\qquad\overset{\overset{O}{\|}}{C}-O-CH_2-\underset{R}{CH}-O^-\,{}^+NHR_3 \qquad (47)$$

VIII

$$VIII \;+\; \overset{O}{\underset{}{C}}\diagdown O \diagup \overset{O}{\underset{}{C}} \longrightarrow A-\overset{\overset{O}{\|}}{C}\quad\overset{\overset{O}{\|}}{C}-O-CH_2-\underset{R}{CH}-O-\overset{\overset{O}{\|}}{C}\quad\overset{\overset{O}{\|}}{C}-O^-\,{}^+NHR_3 \quad (48)$$

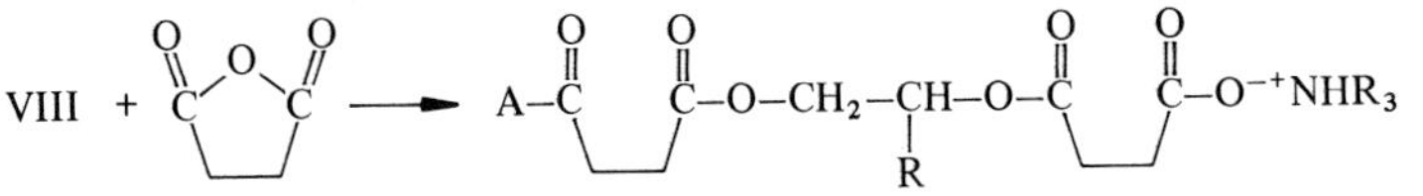

Termination:

$$\sim CH_2-\underset{R}{CH}-O^-\,{}^+NHR_3 \longrightarrow \sim CH_2-\underset{R}{CH}-OH \;+\; NR_3 \qquad (49)$$

Initiation takes place by rapid reaction of an ammonium salt with the anhydride (Eq. (46)) whereby ammonium carboxylate is formed. In the propagation step, the carboxylate anion opens an epoxy ring and forms an ammonium alcoholate (Eq. (47)). The latter reacts with the anhydride to yield another ester bond, and ammonium carboxylate is recovered (Eq. (48)). Termination occurs through decomposition of the ammonium counter ion, the alkoxide anion abstracting a proton from the quaternary nitrogen with the formation of a deactivated tertiary amine.

Tanaka and Kakiuchi [52] proposed a mechanism illustrated by reactions (50)–(54)

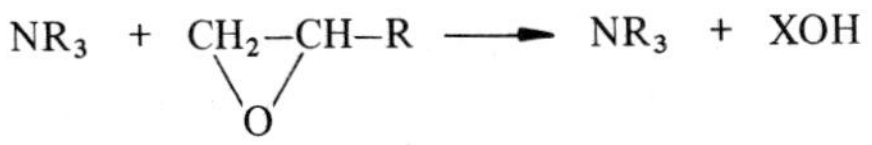

$$NR_3 \;+\; \underset{\underset{O}{\diagdown\diagup}}{CH_2-CH-R} \longrightarrow NR_3 \;+\; XOH \qquad (50)$$

$$NR_3 \;+\; XOH \rightleftharpoons R_3N \ldots HOX \qquad (51)$$

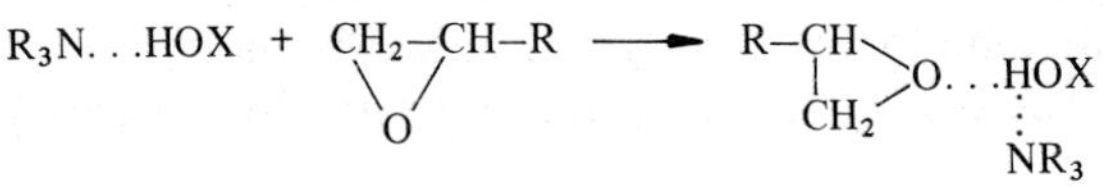

$$R_3N \ldots HOX \;+\; \underset{\underset{O}{\diagdown\diagup}}{CH_2-CH-R} \longrightarrow \underset{CH_2}{\overset{R-CH}{\diagdown}} O \ldots HOX \qquad (52)$$
$$\qquad\qquad\qquad\qquad\qquad\qquad\qquad NR_3$$

IX

$$IX \longrightarrow NR_3 \;+\; XOH$$

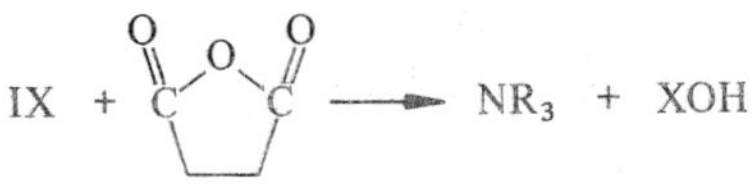

$$IX \;+\; \overset{O}{\underset{}{C}}\diagdown O \diagup \overset{O}{\underset{}{C}} \longrightarrow NR_3 \;+\; XOH \qquad (54)$$

These authors assume in the first step a basically catalyzed isomerization of epoxide yielding an α,β-unsaturated alcohol (Eq. (50)) which forms with the tertiary amine a binary associate through hydrogen bonding (Eq. (51)). This complex forms with the epoxide a ternary activated species IX which is either decomposed according to reaction (53) (with an excess of epoxide formation of polyether sequences) or reacts with the anhydride to generate a monoester (Eq. (54)). The newly formed carboxy group reacts with the tertiary amine to produce a reactive complex (Eq. (51)), and continuation of these alternating steps results in a polyester.

Lustoň et al. [45] combined the mechanisms proposed by Feltzin et al. [73] and Tanaka and Kakiuchi [52] and proposed the following scheme

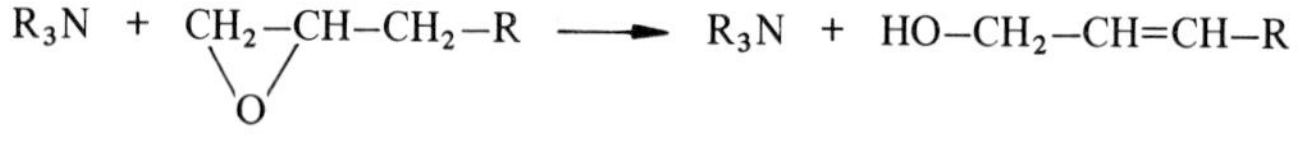

$$R_3N \;+\; CH_2\text{-}CH\text{-}CH_2\text{-}R \longrightarrow R_3N \;+\; HO\text{-}CH_2\text{-}CH=CH\text{-}R \qquad (55)$$
$$\underset{O}{\diagdown\diagup}$$

$$R_3N \;+\; HO\text{-}CH_2\text{-}CH=CH\text{-}R \rightleftharpoons R_3N...HO\text{-}CH_2\text{-}CH=CH\text{-}R \qquad (56)$$
$$X$$

$$X + \;\text{(anhydride)}\; \rightleftharpoons R_3N...H\text{-}O\text{-}CH_2\text{-}CH=CH\text{-}R \qquad (57)$$
$$XI$$

$$XI \longrightarrow R_3HN^{+-}O\text{-}\overset{O}{\overset{\|}{C}}\quad\overset{O}{\overset{\|}{C}}\text{-}O\text{-}CH_2\text{-}CH=CH\text{-}R \qquad (58)$$
$$XII$$

$$XII \;+\; CH_2\text{-}CH\text{-}CH_2\text{-}R \longrightarrow R_3HN^{+-}O\text{-}CH\text{-}CH_2\text{-}O\sim \qquad (59)$$
$$\underset{O}{\diagdown\diagup}\qquad\qquad\qquad\qquad\underset{CH_2\text{-}R}{|}$$
$$XIII$$

$$XIII \;+\; \text{(anhydride)} \longrightarrow R_3HN^{+-}O\text{-}\overset{O}{\overset{\|}{C}}\quad\overset{O}{\overset{\|}{C}}\text{-}O\sim \qquad (60)$$
$$XIV$$

$$XIII \rightleftharpoons R_3N \;+\; HO\text{-}CH\text{-}CH_2\text{-}O\sim \qquad (61)$$
$$\qquad\qquad\qquad\qquad\underset{CH_2\text{-}R}{|}$$

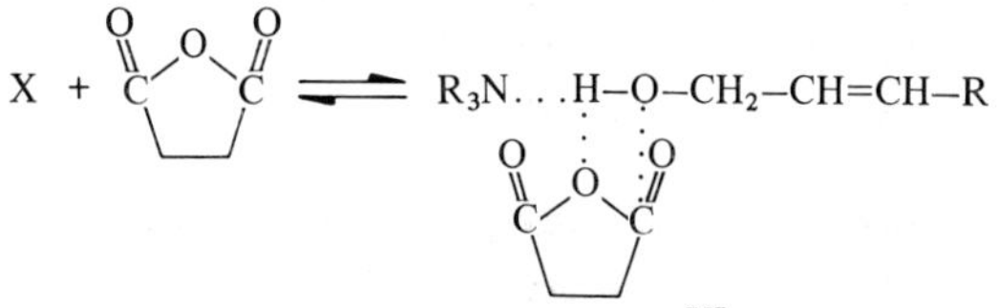

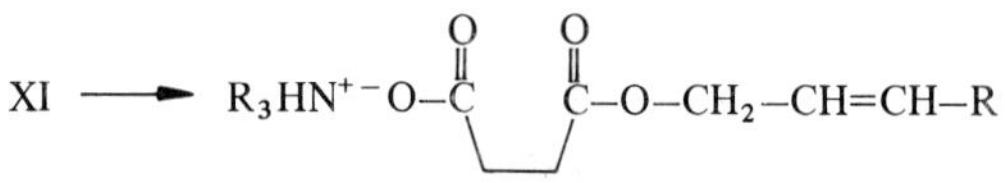

$$XIV \rightleftharpoons R_3N \;+\; HO\text{-}\overset{O}{\overset{\|}{C}}\quad\overset{O}{\overset{\|}{C}}\text{-}O\sim \qquad (62)$$

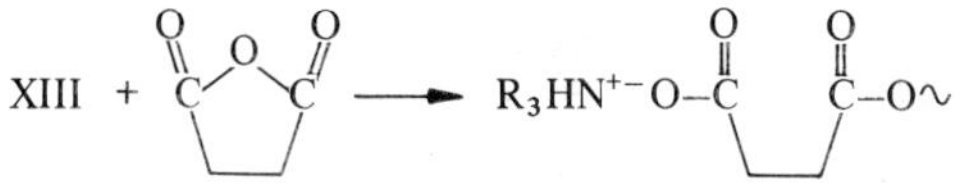

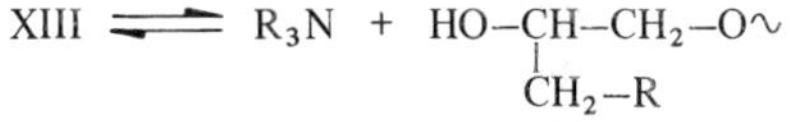

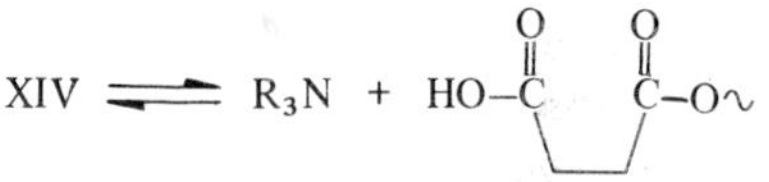

According to this mechanism, the active centre is formed by base-catalyzed isomerization of the epoxide to a substituted alcohol (Eq. (55)). Subsequent reaction of this proton donor with the tertiary amine (Eq. (56)) and anhydride (Eq. (57)) gives a ternary complex which forms an ammonium carboxylate (Eq. (58)). The carboxylate anion of this salt (XII) initiates propagation by reaction with the epoxide yielding the corresponding ammonium alcoholate which again reacts with the anhydride to give the corresponding ammonium carboxylate. Repetition of these steps result in chain growth. Termination proceeds by decomposition of ammonium alcoholate or carboxylate with the release of a tertiary amine and a proton donor which are able to reinitiate the copolymerization process.

For a copolymerization initiated by a tertiary amine in the presence of a proton donor the previous mechanism is modified [74] by introducing a direct interaction of the donor proton with the tertiary amine yielding an active centre. If the proton donor is an acid, the active centre is formed according to

$$C_6H_5COOH + R_3N \rightleftharpoons C_6H_5COOH \ldots NR_3 \rightleftharpoons C_6H_5COO^- \, {}^+NHR_3 \quad (63)$$

If the proton donor is an alcohol or a phenol, the active centre is formed directly in reactions (56)–(58) without epoxide isomerization. The propagation steps are the same as in the previous mechanism (Eqs. (59) and (60)). Antoon and Koenig [67] proposed a copolymerization scheme for the curing of epoxy resins by anhydrides as a refinement of the mechanisms suggested by Tanaka [52] and Lustoň [45, 74]. However, copolymerization again occurs in the presence of proton donors. The complete sequence proposed by Antoon and Koenig [67] is as follows:

$$R_3N + R'OH \rightleftharpoons R_3N \ldots HOR' \quad (64)$$
$$X$$

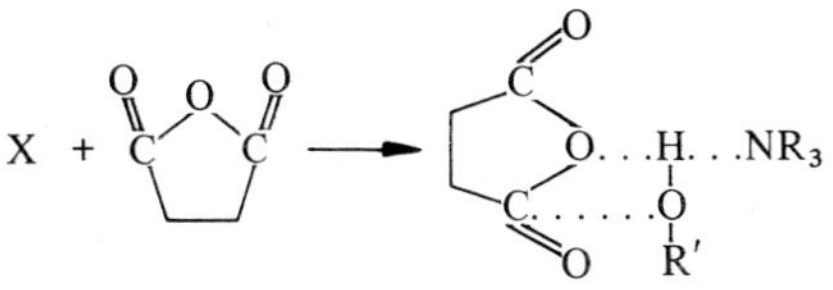

$$(65)$$

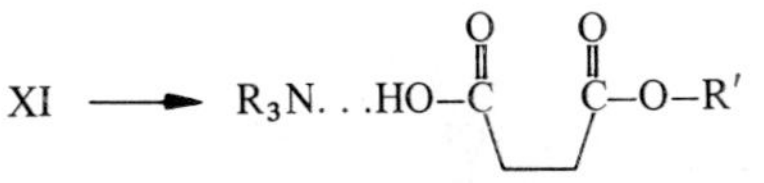

XI

$$(66)$$

XV

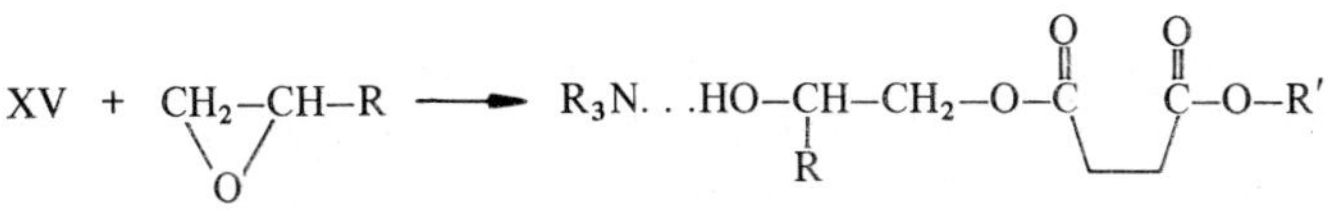

$$(67)$$

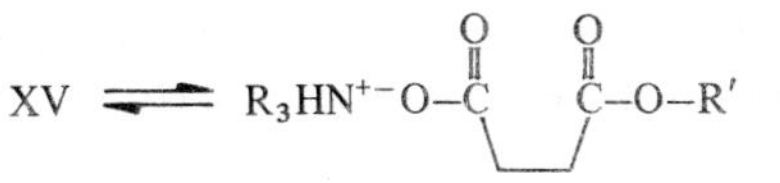

$$(68)$$

Similar as in the previously discussed mechanisms [45, 52, 74], an associate between the tertiary amine and a hydroxy group of the epoxy resin (Eq. (64)) is formed in the first step. In the next step, a ternary complex between adduct X and the anhydride is formed (Eq. (65)) which is converted into the monoester associate with the tertiary amine (XV). The reaction between XV and the epoxide yields a diester with a secondary hydroxy group. The chain growth consists of alternation of reactions (65) and (67). The carboxylate anion (Eq. (68)) is not included in the propagation steps. Fedtke and Mirsojew [74] report that in the system phenylglycidyl ether — benzyl-dimethylamine stilbene and amino alcohol are formed at elevated temperatures, the latter being a source of active hydrogen. Recently, Matějka et al. [68] suggested a new copolymerization scheme in which the polymerization of epoxide is initiated by tertiary amines. For the model reactions of monocarboxylic acid anhydrides with phenylglycidyl ether, the reaction scheme is as follows [68]:

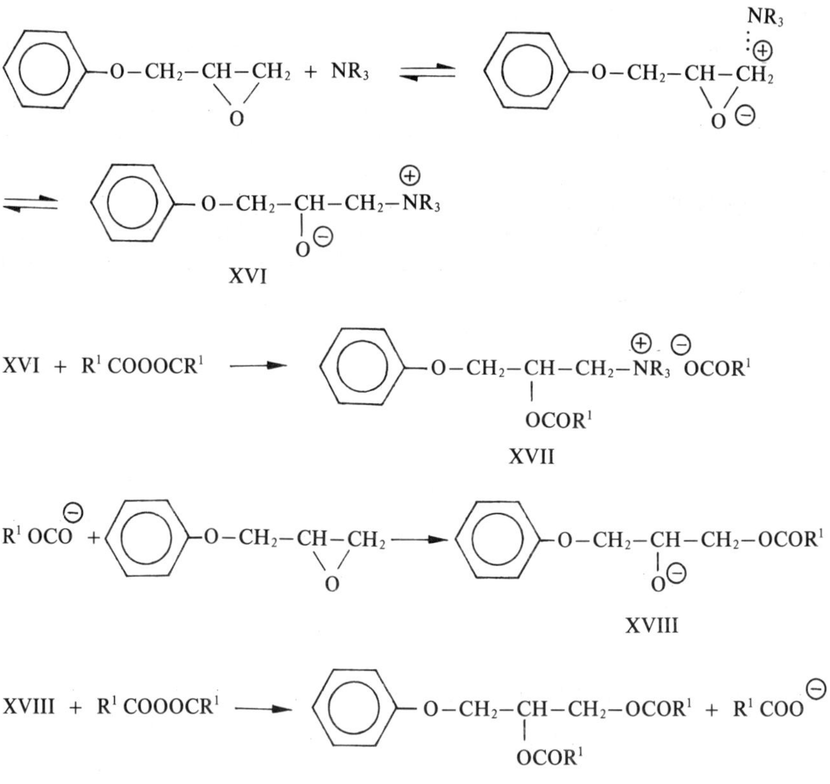

According to this mechanism, initiation includes reaction between the tertiary amine and epoxide, and the primary active centre is represented by a zwitterion with an alkoxide anion and an irreversibly bound amine in the form of an ammonium cation (Eq. (69)). This zwitterion reacts in the next step with the anhydride (Eq. (70)) yielding a carboxylate anion. The growth reactions (Eqs. (71) and (72)) include interactions of the carboxylate anion with epoxide, and of the alkoxide anion with the anhydride.

A comparison of all copolymerization mechanisms shows that there are different interpretations of the nature of the active centre. An anionic copolymerization mechanism has been suggested by Fischer [39, 40], Tanaka and Kakiuchi [36], Feltzin et al. [73], Lustoň and Maňásek [45, 74], and Matějka et al. [68]. In other papers, Tanaka and Kakiuchi [52] and Antoon and Koenig [67] reported the effect of tertiary amines during the formation of transition states. However, the presence of ionic species has recently been proved by conductivity measurements [57], IR [67], and NMR spectroscopy [68]. Although Antoon and Koenig [67] do not consider any activity of carboxylate anions in the propagation reactions, some of their spectral measurements indicate that carboxylate species participate in chain growth. They observed hydrolysis of the anhydride to the corresponding diacid followed by ionization of the carboxy groups by a tertiary amine in the presence of water in the copolymerization system. Only in the early stage of copolymerization can the nonionic carboxy band be detected. The copolymerization with an excess of epoxide shows that, after depletion of the anhydride, the concentration of the carboxylate anions decreases due to reaction with the epoxide. We believe that all these data support the view of the anionic nature of the process.

There are also different views concerning the formation of the initiation centre and its structure. According to Fischer [39, 40], the active centre is directly formed by the reaction of the anhydride with the tertiary amide (Eq. (37)), yielding a betaine-like structure. Tanaka and Kakiuchi [36] modified the initiation scheme by postulating an associate of a tertiary amine with a proton donor. The associate opens the anhydride similarly as in Fischer's mechanism involving formation of a betaine in which the anionic site is stabilized by a hydrogen bond with the original proton donor (Eqs. (41), and (42)). Feltzin et al. [73] proposed an activation of the tertiary amine by the proton donor accompanied by the formation of a quaternary ammonium base (Eq. (45)). The formation of a ternary complex from epoxide, proton-donor compound and tertiary amine as active centre (Eqs. (50–52)) was reported by Tanaka and Kakiuchi [52]. Combination of the last two mechanisms leads to a sequence of reactions (Eqs. (55–58)) where a primary active centre is formed by base-catalyzed epoxide isomerization to allyl alcohols. Its interaction with tertiary amine and anhydride results in the formation of an ammonium salt of a dicarboxylic acid monoester [45]. Antoon and Koenig [67] have suggested the formation of the active centre by reaction of a tertiary amine with a proton donor and subsequently with an anhydride yielding a ternary complex. This complex isomerizes to a dicarboxylic acid monoester. The free carboxy group forms a hydrogen bond with the tertiary amine. Matějka et al. [68] and proposed the formation of an active center by reaction of the tertiary amine with the epoxide (Eq. (69)) yielding a zwitterion with an alkoxide anion and ammonium cation.

The question of initiation, structure and character of the active centre is fundamental for the copolymerization of epoxides with cyclic anhydrides. We therefore analyze arguments supporting individual mechanisms. Since in most mechanisms HA compounds are involved, the question arises whether proton donors are necessary for the formation of an active centre.

Fischer [39, 40] did not observe any effect of allyl alcohol on the copolymerization of allyl glycidyl ether with phthalic anhydride. Reaction rates were identical and indpendent of proton donor concentration. This finding indicates that the presence of

HA compounds is not necessary for the formation of a polyester. Nevertheless, an acceleration effect of HA compounds on the rate of copolymerization was detected later [36, 57, 74], even for systems in which proton donors are directly bound to monomers [67]. This effect is not the sum of the contributions from the tertiary amine and the proton donor but even stronger. Hence, proton donors display a cocatalytic effect. Concerning the effect of HA compounds Tanaka and Kakiuchi [36] established a linear correlation between Hammett's σ constants and the logarithm of the gelation time for various substituted derivatives of benzoic acid, benzyl alcohol and phenol, and positive reaction parameters ϱ were found in all cases. This means that electron-withdrawing substituents increase the effect of HA compounds, or their effect becomes more pronounced with increasing hydrogen atom acidity.

A rise in the content of hydroxyl groups and double bonds was, however, observed during copolymerization without HA compounds by Tanaka and Kakiuchi [52]. These groups and unsaturation can arise by base — or acid-catalyzed isomerization of epoxides to allyl alcohol derivatives [45, 52, 76] (and references cited therein). In the presence of tertiary amines, no isomerization products were isolated probably because of the subsequent epoxide polymerization. However, the presence of unsaturation and hydroxy groups was observed by IR spectroscopy [67, 77, 78] in the polymerization of epoxy resins initiated by tertiary amines. The reaction of trimethylamine with epichlorohydrin described by Burness [79] did not lead to the expected glycidyl trimethylammonium chloride either but yielded its isomer, i.e. N-(3-hydroxy-1-propenyl) trimethylammonium chloride. The reaction proceeds through a transient epoxide which isomerizes according to Eq. (73):

$$R_3N^+-\overset{H}{\underset{H}{C}}-CH-CH_2 \xrightarrow{-B:H} R_3N^+-CH=CH-CH_2-O^- \xrightarrow{H^+}$$

$$\longrightarrow R_3N^+-CH=CH-CH_2-OH \qquad (73)$$

This isomerization is also catalyzed by more sterically hindered amines, and its rate increases with temperature and basicity of the amine used [80]. In the absence of α-hydrogens, the isomerization of epoxide may lead to aldehydes [81]

$$CH_2-CH_2 + R_3N \rightleftharpoons \left[\overset{\delta+}{R_3N}...CH_2-CH_2 \atop \underset{\delta-O}{} \right] \rightleftharpoons$$

$$\rightleftharpoons R_3N + H-\overset{H}{\underset{O^-}{C}}-CH_2^+ \longrightarrow CH_3-CH=O \qquad (74)$$

Antoon and Koenig [67] reject this isomerization of epoxides in copolymerization with anhydrides on the basis of IR spectra. However, in the monomer feed, proton donors are present as hydroxy groups in the epoxy resin and possibly as free carboxy groups in the anhydride. In such a case, no isomerization is necessary. Matějka et al. [68] did not observe any presence of double bonds in the products of the model reaction by ^{1}H-NMR spectroscopy.

Conductivity measurements of the copolymerization system and its binary mixtures showed [57] changes in conductivity during copolymerization as well as the effect of added proton donors. If the rise in the conductivity at the beginning of copolymerization is interpreted as being due to a gradual increase of the concentration of active centres (of course, the conductivity value does not reflect the presence of all active centres but only of free ions in terms of Eq. (14)), a shortening of the time necessary to obtain the maximum on the conductivity curve owing to the presence of HA compounds (Fig. 3) means a shortening of the induction period. Hence, proton donors participate in the formation of active centres.

Proton donors may be present or formed at the beginning of copolymerization and are necessary for the formation of primary active centres. In such a case, initiation according to Fischer's mechanism [39, 40] (Eq. (37)) does not take place. Although the formation of a zwitterion (betaine-like structure) supports the possibility of the generation of charge-transfer complexes between cyclic anhydrides and amines, which may yield betaine-like structures [82], simple anhydrides are much weaker π acceptors than the often employed tetrachlorophthalic anhydride.

According to Menger [83], a simple nucleophilic catalysis is considered to occur in methanolysis of tetrachlorophthalic anhydride in the presence of pyridine, and charge-transfer complex formation has been confirmed neither by kinetic studies nor by spectrometry. Also, the conductivity of a binary solution anhydride-tertiary amine is much lower than that of the ternary system containing an epoxide [57] and does not change with time. Antoon and Koenig [67] also reject the formation of zwitterions. Hence, the first modification of Fischer's mechanism performed by Tanaka and Kakiuchi [36] is not appropriate, which was later admitted by the authors [84]. The formation of ionic species [57, 67, 68] probably proceeds by the anionic mechanism suggested by Feltzin et al. [73] (Eqs. (45) and (46)), Lustoň et al. [45, 74] (Eqs. (55)–(58)) and Matějka et al. [68] (Eq. (69)) rather than by the mechanism assuming the presence of a tertiary amine in a ternary complex [52] (Eqs. (50)–(52)) or binary associates [67] (Eqs. (64) and (66)). The mechanism proposed by Lustoň [45] involves the formation of active centres directly from the components of the copolymerization system and describes copolymerization proceeding in the absence of proton donors. The recently suggested mechanism by Matějka et al. [68] is also assumed to proceed in the absence of proton donors. The initiation reaction (Eq. (69)) has already been suggested for the polymerization of epoxides [85]. However, later, it was observed [86, 87] that homopolymerization of epoxides initiated by tertiary amine did not proceed in the absence of proton donors, and a new initiation mechanism was suggested [86–88] which includes reaction between epoxide, tertiary amine and proton donor yielding an ammonium alcoholate:

$$R^1_3N + R^2-CH-CH_2 + R^3OH \longrightarrow R^2-CH-CH_2-\overset{\oplus}{N}R^1_3 \ \overset{\ominus}{O}R^3 \qquad (75)$$
$$\underset{O}{\diagdown \diagup} \qquad\qquad\qquad \underset{OH}{|}$$

The Feltzin mechanism [73] takes account of the presence of proton donors at the beginning of copolymerization. However, initiation probably proceeds in two ways [74] and depends on the type of the proton donor and its concentration in the copolymerization mixture. If HA in Eq. (45) is alcohol, phenol or moisture, initiation occurs according to Eq. (46), i.e. through interaction with the anhydride yielding an ammonium salt of the monoester. The formation of monoesters as primary active centres accounts here for the lower cocatalytic effect of phenols as compared with alcohols. If the proton donor is a carboxylic acid, activation of the tertiary amine (Eq. (63)) is followed by reaction with the epoxide according to Eq. (76) [74].

$$R-COO^- \; {}^+NHR_3 \; + \; CH_2-CH-R \longrightarrow R_3HN^+ \; {}^-O-CH-CH_2-OOC-R$$
$$\underset{O}{\diagdown\diagup} \qquad\qquad\qquad \underset{R}{|}$$

$$(76)$$

All authors accept the alternating incorporation of epoxide and anhydride into the macromolecular chain [36, 39, 40, 45, 52, 73, 74]. However, the mechanisms of termination and chain transfer have not yet been elucidated. Although the lability of the nitrogen atom is obvious [39, 40, 44] and its salts or associates are readily thermally decomposed [89], Fischer [39] detected its presence in precipitated polyesters by elemental analysis. A simple calculation confirms the presence of the nitrogen atom in almost every tenth macromolecule. In this case, the isolated polyester might be a "living polymer" and, on the addition of monomers, it might initiate another copolymerization. Similar experiments have not been reported so far.

Because the purity of the monomers used affects the molecular weight of the polyester [39], we can assume that proton donors whether added to the monomer mixture or formed during copolymerization, participate in termination or transfer reactions. Termination of chain growth may occur by reaction of the growing chain end with proton donors according to Eqs. (77) and (78).

$$\sim CH_2-O^- \; {}^+Cat + Ha \; \rightarrow \; \sim CH_2-OH + CatA \qquad (77)$$

$$\sim COO^- \; {}^+Cat + HA \rightarrow \; \sim COOH + CatA \qquad (78)$$

In these processes hydroxy and carboxy end groups and salts capable to reinitiate copolymerization are generated. New end groups can again be formed in termination reactions. In fact, Eqs. (77) and (78) describe a chain transfer. This is also supported by the fact that during copolymerization about 10 chains are formed per one molecule of the tertiary amine at equimolar epoxide — to — anhydride ratio [39, 44].

3.3.4 Kinetics of Copolymerization

The kinetics of copolymerization or curing of epoxy resins with cyclic anhydrides initiated by tertiary amines was investigated by chemical analysis [32, 35, 36, 39, 40, 45, 52, 65, 73, 74, 90], differential scanning calorimetry [91-94], isothermal methods [95], electric methods [96], dynamic differential thermal analysis [97], IR spectroscopy [45, 65, 67], dilatometry [90], or viscometry [90]. Results of kinetic measurements and their interpretation differ; most authors agree, however, that the copolymerization is of first order with respect to the tertiary amine.

Great differences in evaluating kinetic measurements and in determining the order of copolymerization are due to high copolymerization rates in the melt and to a relatively low precision in determining the conversion curves, especially at low conversions. In such cases, the copolymerization rate seems to be constant and independent of monomer concentration.

Kinetic curves obtained for a sufficiently slow reaction (low initiator concentration, low temperature, copolymerization in solution) show a sigmoidal shape [35, 36, 45, 52, 74] (Fig. 6) with an induction period.

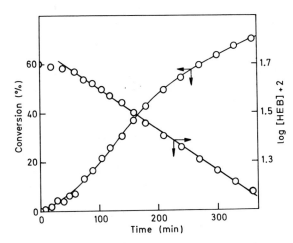

Fig. 6. Copolymerization of 2-hydroxy-4-(2,3-epoxypropoxy)-benzophenone (HEB) (0.5 mol/l) with phthalic anhydride (0.5 mol/l) in o-xylene at 120 °C initiated by tri-n-hexylamine (0.025 mol/l) [45]

Clear indications of the induction period and of an increase in the reaction rate after copolymerization has started were found for isothermal runs by DSC measurements by Peyser and Bascom [94] even for melt copolymerization. According to the copolymerization mechanism, the induction period is interpreted as a gradual increase in the concentration of active centres [45, 52] and is identical with the time for reaching the maximum on the conductivity curves [57]. An induction period has also been established by other measurements [39, 40, 73, 90, 95], where it is often considered as an imprecision in the determination of the monomer concentration, mixing effect, temperature establishement, or it is not considered at all.

The induction period is followed by the region of the maximum rate of copolymerization [52]. Conversion curves can be correlated in this region [45, 74] with the first-order kinetic equation (Fig. 6) and the rate of copolymerization is thus of the first oder with respect to one of the monomers. Also, recalculation of some experimental data found in the literature [35, 52, 90] gives a satisfactory fit to the first order kinetic equation. If the rate of copolymerization is first-order with respect to the tertiary amine [19, 32, 35, 36, 39, 40, 45, 52, 65, 67, 73, 74, 95, 97, 98], the overall rate of copolymerization is of second order.

A reaction order of 1/2 was determined by Malavašič et al. [91] for the curing of epoxy resins with cyclic anhydrides over the conversion range 18–79%. At 86–98.5% conversion, the authors established a first-order curing reaction. Booss and Hauschildt [90] regard copolymerization and curing as a zero-order process with respect to anhydride or epoxide and consider the reaction order to be dependent on temperature and conversion. Other authors like Fischer [39,40], Arnold [19], Feltzin et al. [73], Yaralov and Obanesova [97] and Taratorin and Alekseeva [95] confirmed first-order kinetics with respect to the amine for curing of resins and copolymerization and observed zero-order kinetics with respect to epoxide and anhydride.

Feltzin et al. [73] have reported the first-order kinetics and express the rate constant by the Arrhenius equation. Thus, for the curing of diglycidyl ether of Bisphenol A with nadic methyl anhydride, initiated by benzyldimethylamine, we have[3]:

$$-\frac{d\,[\text{epoxide}]}{dt} = 2.5 \times 10^{-5}\,[\text{initiator}]\,\exp\,(-27\,600/RT) \tag{79}$$

Tanaka and Kakiuchi [52] analyzed the conversion curves for the copolymerization of substituted phenylglycidyl ethers with hexahydrophthalic anhydride and found a first-order dependence with respect to the initiator at the maximum reaction rate (Eq. (80)), and second-order kinetics for the region of initiation (Eq. (81)).

$$v = k\,[\text{tert. amine}]_0 \tag{80}$$

$$v = k[\text{epoxide}]_0\,[\text{tert. amine}]_0 \tag{81}$$

where $[\text{epoxide}]_0$ and $[\text{tert. amine}]_0$ are the initial concentrations of the epoxide and amine.

Antoon and Koenig [67] found a first-order dependence with respect to the tertiary amine, but they were not able to distinguish between zero-, first-, and second-order kinetics with respect to monomers. Approximately the same fit was obtained for all mentioned orders below the gel point. Peyser and Bascom reported first-order kinetics for the curing of the diglycidyl ether of bisphenol A with hexahydrophthalic anhydride [94] and second-order kinetics with respect to the same epoxide cured with nadic methyl anhydride [93]. They also pointed out the uncertainty in determining the reaction order.

Second-order kinetics with respect to the amine and epoxide was also found by Antipova et al. [65] for the curing of epoxy resins with hexahydrophthalic anhydride, by Sorokin et al. [32] for the reaction of phenylglycidyl ether with phthalic anhydride in the presence of butanol, Lustoň and Maňásek [45,74] for the copolymerization of 2-hydroxy-4-(2,3-epoxypropoxy)benzophenone with phthalic anhydride in the absence or in the presence of proton donors, and Kudyakov et al. [98] for the curing of epoxy resins with maleic anhydride.

Tanaka and Kakiuchi [35,36] found a proportionality between the rate of curing of epoxy resins with hexahydrophthalic anhydride and the concentrations of epoxide, anhydride and tertiary amine (Eq. (82)), and also first-order kinetics with respect to the proton donors if present (Eq. (83)).

$$v = k\,[\text{epoxide}]\,[\text{anhydride}]\,[\text{tert. amine}] \tag{82}$$

$$v = k\,[\text{epoxide}]\,[\text{anhydride}]\,[\text{tert. amine}]\,[\text{proton-donor}] \tag{83}$$

---

[3] Recalculation of the authors' experimental results gives the activation energy $E_A = 55.7$ kJ/mol

The copolymerization mechanisms show that the propagation reactions are bimolecular in two alternating steps and that the rate-determining step is the slower propagation reaction. In our case, the slower reaction is the addition of the epoxide (Eqs. (38), (43), (47), (52), (59), (67), and (71) in the respective schemes) [99] as has also been found for the initiation by ammonium salts [56]. Since the tertiary amine does not directly take part in growth reactions (cf. 3.3.3), a more suitable expression for the copolymerization rate is Eq. (84) where the tertiary amine should be replaced by an active centre.

$$v = -\frac{d\,[\text{epoxide}]}{dt} = k\,[\text{epoxide}]\,[\text{active centre}] \qquad (84)$$

It has been shown experimentally [68,74] that the active centres are formed at the beginning of copolymerization during the induction period; their concentration remains constant throughout the whole copolymerization period (steady state). Although Antoon and Koenig [67] observed an increase in the hydroxyl concentration during copolymerization, which was interpreted as an argument against the establishement of a steady state, they also observed a relatively constant intensity of the carboxylate absorption band. However, in the kinetic scheme they established they also assume a steady-state.

Any group capable of inducing the addition of monomers, i.e. for the mechanism described e.g. by Eqs. (69)–(72) the zwitterion as well as the carboxylate and alkoxide anion is considered as an active centre.

Since the concentration of these active centres increases in the initial stage of copolymerization where only small changes in the concentration of the components occur, initial monomer concentrations can be used in the kinetic equation for the formation of the active centres. The concentration of active centres is usually much lower than the concentration of any other component participating in the equilibrium. Since reaction (55) has not been confirmed experimentally and is assumed to be rather slow under the given experimental conditions, the existence of the induction period followed by a steady state is compatible with the scheme described by Eqs. (69)–(72). For initiation without proton-donor compounds it holds [45]:

$$[\text{active centre}] = K_c\,[\text{epoxide}]_0\,[\text{anhydride}]_0\,[\text{tert. amine}]_0 \qquad (85)$$

where K_c is the complex equilibrium constant for the formation of an active centre.

By introducing Eq. (85) into Eq. (84), we obtain for the rate of copolymerization

$$v = -\frac{d\,[\text{epoxide}]}{dt} = k_c\,[\text{epoxide}]_0\,[\text{anhydride}]_0\,[\text{tert. amino}]_0\,[\text{epoxide}]$$
$$(86)$$

The copolymerization of epoxides with cyclic anhydrides is a thermally activated reaction. Table 6 gives a survey of the thermodynamic parameters. The activation energies determined by different authors are in good agreement and vary between 52.8 and 64.9 kJ/mol, depending on the monomer used. Exceptions are only the

Table 6. Kinetic and thermodynamic parameters for epoxide-anhydride-initiator copolymerization[a]

Epoxide	Anhydride	Initiator	Overall reaction order	E_a (kJ/mol)	log A	S^*_{298} (e. u.)	H^*_{298} (kJ/mol)	G^*_{298} (kJ/mol)	Ref.
PGE	HHPA	DMBA	0	95.9[b]	13.7				90)
				27.0[c]	5.5				
				46–54[d]					
Rütapox 0180 E	PA	HMTA	0	86.4	11.8				90)
Rütapox 0162	HHPA	DMBA	0	86.4	11.8				90)
Araldit CY 205	X 157/2378	DY 062	1/2	58.2[e]					91)
			1	56.1[f]					
DER 332	NMA	DMBA	1	27.6	−6.4				73)
				55.7[g]	5.5[g]				
DER 332	HHPA	DMBA	1	104.7[h]	27				94)
				58.6[i]					
				88.8[j]	20				
ED-5	MA	DMA	1	100					95)
ED-5	MTHPA	DMA	1	100					95)
ED-5	MA	DMA	1	211					97)
ED-5	PA	DMA	1	98					97)
ED-5	MTHPA	DMA	1	172					97)
ED-5	THPA	DMA	1	144					97)
ED-5	CA	DMA	1	106.3					97)
ED-5	SA	DMA	1	105.9					97)
ED-5	GA	DMA	1	119.7					97)
p-NO$_2$ PGE	HHPA	TBA	2	56.9[k]	3.9	−44.1	51.9	106.8	52)
			1	52.8[l]	4.1	−43.9	47.7	102.6	
p-CO$_2$CH$_3$ PGE	HHPA	TBA	2	58.2[k]	4.0	−44.1	53.2	108.0	52)
			1	53.2[l]	4.0	−43.8	48.1	102.6	
p-Cl PGE	HHPA	TBA	2	59.0[k]	4.1	−43.6	54.0	108.4	52)
			1	53.6[l]	4.0	−43.8	48.6	103.0	
PGE	HHPA	TBA	2	60.3[k]	4.2	−43.3	55.3	109.3	52)
			1	53.6[l]	3.9	−44.3	48.6	103.8	

Epoxide	Anhydride	Amine	n						Ref.
p-CH$_3$ PGE	HHPA	TBA	2	61.5[k] / 54.0[l]	4.3 / 4.0	−42.8 / −44.1	56.5 / 49.0	110.1 / 103.8	52)
p-OCH$_3$ PGE	HHPA	TBA	1	62.0[k] / 54.0[l]	4.3 / 4.0	−42.6 / −44.4	56.9 / 49.0	110.1 / 104.3	52)
DER 332	NMA	DMBA	1–2	55.3–63.2[m] / 147.4–177.1[n]	10.5–11.1[m] / 38–46[n]				93)
HEB	PA	THA	2	58.2	3.6				45)
Epikote 828	HHPA	TEA	2	59.0 / 57.4[d]					35)
Epikote 1001	HHPA	TEA	3	60.3	6.7	−35.0			35)
Epikote 828	HHPA	TEA	3	59.0	6.8	−34.8			36)
Epikote 1001	HHPA	TEA	3	60.3	6.4	−35.0			36)
Epikote 828	NMA	TEA	4	61.5[o]	6.2	−35.4			36)
Epikote 828	NMA	TEA	4	61.5[p]	6.3	−35.2			36)
Epikote 828	NMA	TEA	4	61.5[q]					36)
DER 332	HHPA	TAP	?	74.5					92)
Epon 828	NMA	DMBA	?	64.9					67)

^a For symbols cf. List of abbreviations;
^b up to 75% conversion;
^c over 75% conversion;
^d from viscosity measurements;
^e between 18 and 80% conversion;
^f over 80% conversion;
^g recalculated values according to experimental results of Feltzin [73];
^h from DSC dynamic measurements;
ⁱ from DSC isothermal measurements (reaction order n = 1);
^j from DSC isothermal measurements (taking into account changes in the reaction order;
^k for initiation range;
^l for final range;
^m for range of maximum rate;
ⁿ for initial region;
^o in the presence of benzoic acid;
^p in the presence of benzyl alcohol;
^q in the presence of phenol

results of Feltzin et al. [73], who determined lower activation energy, and of Teratorin [95], Yaralov [97], Peyser and Bascom [93,94], and Evtuschenko [96], who found much higher activation energies. Recalculation of the experimental data of Feltzin [73] gives the activation energy $E_A = 55.7$ kJ/mol which is in a good agreement with the results of other authors. Differences in the results of Peyser and Bascom [93,94], Taratorin [95], Evtuschenko [96], and Yaralov [97] may depend on the way of processing the thermoanalytical data [93,94,100].

Boos and Hauschildt [90] obtained for the model copolymerization of phenylglycidyl ether with hexahydrophthalic anhydride activation energies of 96 kJ/mol up to 75 % conversion and 27 kJ/mol for higher conversions. Frequency factors are also very different (log A = 13.7 and 5.5, respectively). The frequency factors as well as the temperature coefficients of the solution viscosities depended on the initiator concentration. The activation energy determined by the same authors [90] for the curing of epoxy resins at conversions lower than 75 % was 86.4 kJ/mol and the frequency factor log A = 11.8 whereas at higher conversions these values were not obtained.

4 Conclusion

This review attempts to compile data on the anionic copolymerization of cyclic ethers with cyclic anhydrides. For a better understanding we have also included data and theoretical views on non-catalyzed reactions.

Although this copolymerization has found application as a curing reaction for technically important epoxy resins, gel formation and difficulties connected with the evaluation of chemical analysis data have often led to different interpretations of the results. Attention has often been concentrated on kinetic problems of the copolymerization, but its mechanism has not yet been solved satisfactorily.

At present, we can say that copolymerization initiated by various salts proceeds by an anionic mechanism, after dissociation of the initiators in the reaction medium. The primary step is the addition of the initiator anion to the epoxide. In the initiation by Lewis bases, i.e. by tertiary amines, initiation involves formation of a primary active centre of an anionic character. This active centre is probably generated by interaction of the tertiary amine with the anhydride and an allyl alcohol. The allyl alcohol can be formed by a base-catalyzed isomerization of the epoxide. In the presence of a proton donor, the formation of active centres is possible through interaction of tertiary amine, anhydride and proton donor without epoxide isomerization. Another way of initiation consists in a direct reaction of epoxide with tertiary amine yielding an anionic primary active centre. We believe that in both kinds of initiation in the strict absence of proton donors, the growing chain end has the character of a "living" polymer. The presence of proton donors, however, gives rise to transfer reactions.

In our opinion, the area of this type of copolymerization is still open to discussion and research, and this suggestion concerns mainly the mechanism of copolymerization, formation of active centres in the initiation by Lewis bases, the influence of proton donors on the course of copolymerization, and the effect of the structure of the initiator on the rate of copolymerization. It is necessary, however, to study the copolymerization on model compounds.

5 References

1. Weinschenk, A.: Chem. Ztg. *29*, 1311 (1905)
2. Nishikubo, T., Imaura, M., Takaoka, T.: Nippon Kagaku Kaishi *1974*, 185
3. Hilt, A., Hamann, K., Keifer, S.: Ger. Pat. 1 570 411 (1965)
4. Nollen, K., Kaden, V., Hamann, K.: Angew. Makromol. Chem. *6*, 1 (1969)
5. Kroker, R., Hamann, K.: Angew. Makromol. Chem. *13*, 1 (1970)
6. Lustoň, J., Guniš, J., Maňásek, Z.: J. Macromol. Sci.-Chem. *A7*, 587 (1973)
7. Lustoň, J., Schubertová, N., Maňásek, Z.: J. Polym. Sci., Symp. No. *40*, 33 (1973)
8. Nishikubo, T. et al.: Nippon Kagaku Kaishi *1973*, 1851
9. Matsuda, H.: J. Polym. Sci., Polym. Chem. Ed. *14*, 1783 (1976)
10. Matsuda, H.: J. Appl. Polym. Sci. *22*, 2093 (1978)
11. Matsuda, H., Minoura, Y.: J. Appl. Polym. Sci. *24*, 811 (1979)
12. Matsuda, H.: J. Appl. Polym. Sci. *25*, 1915 (1980)
13. Matsuda, H.: J. Appl. Polym. Sci. *25*, 2339 (1980)
14. Lee, H., Neville, K.: Handbook of Epoxy Resins. New York: McGraw Hill 1967
15. Sorokin, M. F., Shode, L. G.: Lakokrasoch. Mater. Ikh Primen. *1968*, 76; C. A. *68*, 87768 (1968)
16. Ishii, Y., Sakai, S., in: Ring Opening Polymerization. Frisch, K. C., Reegen, S. L. (eds.). New York: Marcel Dekker 1968
17. Tanaka, Y., Mika, T. F., in: Epoxy Resins. May, C. A., Tanaka, Y. (eds.). New York: Marcel Dekker 1973
18. Reichert, K.-H. W.: Chimia *29*, 453 (1975)
19. Arnold, R. J.: Mod. Plast. **41** (8), 149 (1964)
20. Fisch, W., Hofmann, W.: Makromol. Chem. *44*, 8 (1961)
21. Kaplan, S. L., McAllister, L. E., Stewart, A. T.: Polym. Eng. Sci. *6*, 65 (1966)
22. Fisch, W., Hofmann, W.: J. Polym. Sci. *12*, 497 (1954)
23. Fisch, W., Hofmann, W., Koskikallio, J.: J. Appl. Chem. *6*, 429 (1956)
24. Fisch, W., Hofmann, W., Koskikallio, J.: Chem. Ind. (London) *1956*, 756
25. Dearborn, E. C., Fuoss, R. M., White, A. F.: J. Polym. Sci. *16*, 201 (1955)
26. Schechter, L., Wynstra, J.: Ind. Eng. Chem. *48*, 86 (1956)
27. Weiss, H. K.: Ind. Eng. Chem. *49*, 1089 (1957)
28. Shimazaki, A., Kozima, M.: J. Chem. Soc. Jpn., Ind. Chem. Sec. (Kogyo Kagaku Zashi) *66*, 1610 (1963)
29. Shimazaki, A.: J. Chem. Soc. Jpn., Ind. Chem. Sec. (Kogyo Kagaku Zashi) *67*, 1304 (1964)
30. Kannebley, G.: Kunststoffe *47*, 693 (1957)
31. Sorokin, M. F., Shode, L. G., Gershanova, E. L.: Lakokrasoch. Mater. Poluprod. *1967* (5), 67; C.A. *68*, 96461 (1968)
32. Sorokin, M. F., Shode, L. G., Gershanova, E. L.: Chem. Prum. *17*/42, 590 (1967)
33. Schechter, L., Wynstra, J., Kurkjy, R. P.: Ind. Eng. Chem. *49*, 1107 (1957)
34. Pirozhnaya, L. M.: Plast. Massy *1961* (6), 56
35. Tanaka, Y., Kakiuchi, H.: J. Appl. Polym. Sci. *7*, 1063 (1963)
36. Tanaka, Y., Kakiuchi, H.: J. Polym. Sci. *A2*, 3405 (1964)
37. Doszlop, S., Vargha, V., Horkay, F.: Periodica Polytech. (Budapest) *22*, 253 (1978)
38. Patat, F., Wojtech, B.: Makromol. Chem. *37*, 1 (1960)
39. Fischer, R. F.: J. Polym. Sci. *44*, 155 (1960)
40. Fischer, R. F.: Ind. Eng. Chem. *52*, 321 (1960)
41. Schwenk, E. et al.: Makromol. Chem. *51*, 53 (1962)
42. Hilt, A., Reichert, K.-H. W., Hamann, K.: Makromol. Chem. *101*, 246 (1967)
43. Šňupárek, J., Mleziva, J.: Chem. Prum. *18*/43, 473 (1968)
44. Tanaka, Y., Haung, C. M.: Makromol. Chem. *120*, 1 (1968)
45. Lustoň, J., Maňásek, Z., Kulíčková, M.: J. Macromol. Sci.-Chem. *A12*, 995 (1978)
46. Dreyfuss, P., Dreyfuss, M. P., in: Ring-Opening Polymerization. Frisch, K. C., Reegen, S. L. (eds.), p. 112. New York: Marcel Dekker 1968
47. Ref. 16, p. 17
48. Ref. 17, p. 13

49. Nelson, R. A., Jessup, R. S.: J. Res. Natl. Bur. Std. *48*, 206 (1952)
50. Chapman, N. B., Isaacs, N. S., Parker, R. E.: J. Chem. Soc. *1959*, 1925
51. Parker, R. E., Isaacs, N. S.: Chem. Rev. *59*, 737 (1959)
52. Tanaka, Y., Kakiuchi, H.: J. Macromol. Chem. *1*, 307 (1966)
53. Tsirkin, M. Z., Molotkov, R. V., Kazanskaia, V. F.: Plast. Massy *1963* (7), 17
54. Hilt, A., Trivedi, J., Hamann, K.: Makromol. Chem. *89*, 177 (1965)
55. Hilt, A., Hamann, K.: Makromol. Chem. *92*, 55 (1966)
56. Luston, J., Maňásek, Z.: Makromol. Chem. *181*, 545 (1980)
57. Luston, J., Maňásek, Z.: J. Macromol. Sci. Chem. *A12*, 983 (1978)
58. Klaban, J., Smrčka, J., Mleziva, J.: Makromol. Chem. *111*, 1 (1968)
59. Dobinson, B., Hofmann, W., Stark, B. P.: The Determination of Epoxide Groups, p. 18. London: Pergamon Press 1969
60. Szwarc, M.: Carbanions, Living Polymers and Electron Transfer Processes, Chap. VII. New York: Interscience Publishers 1968
61. Herold, B. J.: Catal. Rev., Sci. Eng. *17*, 1 (1978)
62. Kazanskii, K. S.: Main lecture at the 21st Prague Microsymposium on Macromolecules, Ring-Opening Polymerization of Heterocycles, Karlovy Vary 1980, Pure Appl. Chem. *55*, 1645 (1981)
63. Smith, J. D. B.: J. Appl. Polym. Sci. *23*, 1385 (1979)
64. Červinka, O., Dědek, V., Ferles, M.: Organická chemie (Organic Chemistry), p. 609. Prague: SNTL 1969
65. Antipova, L. M. et al.: Plast. Massy *1973*, 12
66. Mayahi, M. F., El-Bermani, M. F.: Can. J. Chem. *51*, 3539 (1973)
67. Antoon, M. K., Koenig, J. L.: J. Polym. Sci., Polym. Chem. Ed. *19*, 549 (1981)
68. Matějka, L. et al.: J. Polym. Sci., Polym. Chem. Ed. 21, (1983)
69. Luston, J., Maňásek, Z.: React. Kinet. Catal. Lett. *9*, 47 (1978)
70. Alvey, F. B.: J. Appl. Polym. Sci. *13*, 1473 (1969)
71. Bogatkov, S. V., Popov, A. F., Litvinenko, L. M.: Reakts. Sposobn. Org. Soedin. *6*, 1011 (1969)
72. Bogatkov, S. V., Zaslavskii, V. G., Litvinenko, L. M.: Dokl. Akad. Nauk SSSR *210*, 97 (1973)
73. Feltzin, J. et al.: J. Macromol. Sci.-Chem. *A3*, 261 (1969)
74. Luston, J., Maňásek, Z.: J. Macromol. Sci.-Chem. *A13*, 853 (1979)
75. Fedtke, M., Mirsojew, I.: Plaste und Kautschuk *28*, 369 (1981)
76. Yandovskii, V. N., Ershov, B. A.: Usp. Khim. *41*, 785 (1972)
77. Sorokin, M. F., Shode, L. G., Steinpress, A. B.: Vysokomol. Soedin. *A13*, 747 (1971)
78. Pirozhnaya, L. N.: Vysokomol. Soedin. *A12*, 2446 (1970)
79. Burness, D. M.: J. Org. Chem. *29*, 1862 (1964)
80. McClure, J. D.: J. Org. Chem. *35*, 2059 (1970)
81. Katsura, I., Kawaguchi, H., Yamamoto, T.: Nippon Kagaku Kaishi *1973*, 1733
82. Nagy, J. B., Bruylants, A., Nagy, O. B.: Tetrahedron Lett. *54*, 4825 (1969)
83. Menger, F. M.: J. Am. Chem. Soc. *90*, 4378 (1968)
84. Ref. 17, p. 190
85. Narracot, E. S.: Brit. Plastics *26*, 120 (1953)
86. Shechter, L., Wynstra, J.: Ind. Eng. Chem. *48*, 86 (1956)
87. Kushch, P. P., Komarov, B. A., Rozenberg, B. A.: Vysokomol. Soedin. *A21*, 1697 (1979)
88. Sorokin, M. F., Shode, L. G., Steinpress, A. B.: Vysokomol. Soedin. *B11*, 172 (1969)
89. Harlow, G. A.: Anal. Chem. *34*, 1487 (1962)
90. Boos, H. J., Hauschildt, K. R.: Angew. Makromol. Chem. *84*, 51, (1980)
91. Malavašič, T., et al.: Angew. Makromol. Chem. 44, *89* (1975)
92. Fava, R. A.: Polymer *9*, 137 (1968)
93. Peyser, P., Bascom, W. D.: Analytical Calorimetry, Porter, R. S., Johnson, J. F. (eds.), Vol. 3, p. 537. New York: Plenum Press 1974
94. Peyser, P., Bascom, W. D.: J. Appl. Polym. Sci. *21*, 2359 (1977)
95. Taratorin, B. I., Alekseeva, N. N.: Vysokomol. Soedin. *A10*, 2569 (1968)
96. Evtushenko, G. T., Moshinskii, L. Ya., Beletskaya, T. V.: Vysokomol. Soedin. *A16*, 1343 (1974)

97. Yaralov, L. K., Obanesova, G. S.: Izv. Vysch. Ucheb. Zaved. Khim. Khim. Tekhnol. *18*, 1571 (1975), C.A. *84*, 60394 (1976)
98. Kudyukov, Yu. P. et al.: Khim. Khim. Tekhnol. Sint. Issled. Plenkoobraz. Veshtchestv. Pigm. *1976*, 32; C.A. *89*, 130229 (1978)
99. Ref. 16, p. 94
100. Kamon, T. et al.: Kogyo Kagaku Zashi *72*, 2677 (1969)

Received April 6, 1983
Prof. K. Dušek (Editor)

Author Index Volumes 1–56

Subject Index

M. Szwarc

Living Polymers and Mechanisms of Anionic Polymerization

1983. 69 figures. V, 187 pages. (Advances in Polymer Science, Volume 49). ISBN 3-540-12047-5

Contents: Introduction. – Thermodynamics of Polymerization. – Initiation of Anionic Polymerization. – Propagation of Anionic Polymerization. – Concluding Remarks. – References.

Michael Szwarc is the "father" of the term "living polymers" which he proposed for those macromolecules that may spontaneously resume their growth whenever fresh monomer is supplied to the system. While the conventional polymerization scheme created the impression that a terminationless polymerization is highly improbable, Szwarc and his associates demonstrated the terminationless character of anionic polymerization of vinyl monomers in the absence of impurities around 1956. Living polymers, although not named in this way, were described earlier by Ziegler. However, Szwarc was able to reveal their characteristic features: they do not die but remain acitve waiting for the next monomer. If the monomer added is different from the one previously used, a "block polymer" results. This, indeed, is the most versatile technique for synthesizing block polymers.

In this review, the author describes the mechanisms of anionic polymerization and pays special attention to the concept and application of living polymers. The role of various species, e.g. free ions, ion-pairs, triple ions etc. is stressed and their meaning is clarified. Their nature and the pertinent interrelations, both thermodynamic and kinetic, are explained. The article provides a background to anybody interested in organic ion reactions and ionic polymerization. (487 ref.)

Springer-Verlag
Berlin
Heidelberg
New York
Tokyo

A. Gandini, H. Cheradame

Cationic Polymerisation

Initiation Processes with Alkenyl Monomers

1980. 12 figures, 9 tables. X, 289 pages. (Advances in Polymer Science, Volume 34/35). ISBN 3-540-10049-0

Contents: Introduction. – Fundamentals. – Initiation by Brønsted Acids and Iodine. – Initiation by Lewis Acids. – Initiation by Carbenium Salts and Related Species. – Initiation by Bare Cations. – Electrochemical Initiation. – Photoinitiation. – Initiation from a Charge-Transfer Complex. – Initiation from a Polymer. – Miscellaneous Initiators. – References. – Subject Index.

This monograph covers the entire spectrum of initiation systems in the cationic polymerisation of alkenyl monomers. Following a detailed outline of the factors which play an important role in determining the behaviour of cationic polymerisation, each type of initiation is discussed individually. Particular emphasis is placed on the two major modes of initiation: initiation by Brønsted acids and initiation by Lewis acids. The authors analyze the present status of this discipline through a critical review of the literature and a series of specific mechanistic proposals, some of which are entirely new. Published material relevant to the understanding of the processes leading to the formation and characterisation of active species is covered exhaustively. The significance of early work is reinterpreted and the impact of more recent studies as well as their shortcomings assessed. The potentials of new experimental techniques are also discussed. Finally, suggestions are offered for future work in many areas on the basis of the mechanistic proposals developed.
This book will help stimulate further ideas, discussions and research in a discipline which is experiencing a lively renaissance.

S. Penczek, P. Kubisa, K. Matyjaszewski

Cationic Ring-Opening Polymerization of Heterocyclic Monomers

1980. 19 figures. V, 156 pages (Advances in Polymer Science, Volume 37) ISBN 3-540-10209-4

Contents: Introduction. – Monomer Structures, Ring Strains and Nucleophilicities (Basicities). – Initiation. – Propagation. – Termination and Transfer Processes. – Addendum. – References. – Subject Index.

The detailied understanding of the chemistry of the elementary Reactions of cationic polymerization, including the corresponding kinetic parameters, is only available for the ring-opening polymerization of heterocyclics. In this volume, the authors present a modern review of cationic polymerization based on the results obtained primarily during the last decade. Although discussion is centered mainly on the polymerization of heterocyclic monomers, some general rules are formulated. Elementary reactions are treated separately; quantitative data were carefully selected to present the relationship between structure and reactivity. For deeper insights into the problems related to cationic polymerization, this review is organized such that similar **mechanisms** rather than similar groups of monomers are trested. The ways in which cationic polymerization of heterocyclic monomers can be initiated, the mechanisms of growth of polymer chain, involving related equilibria and reactions responsible for termination of material and kinetic chain, are analysed from the point of view of their similarities and differences. The review not only summarises the present state of knowledge but also indicates future trends in studies of cationic polymerization.

Springer-Verlag Berlin Heidelberg New York Tokyo